Nature's Mind

OTHER BOOKS BY MICHAEL S. GAZZANIGA

The Bisected Brain

The Integrated Mind (with Joseph LeDoux)

The Social Brain

Mind Matters

Nature's Mind

The Biological Roots of
Thinking, Emotions,
Sexuality, Language, and
Intelligence

Michael S. Gazzaniga, Ph. D.

BasicBooks
A Division of HarperCollins*Publishers*

Library of Congress Cataloging-in-Publication Data
Gazzaniga, Michael S.
    Nature's mind: the biological roots of thinking, emotions, sexuality, language, and intelligence.
        p.   cm.
    Includes bibliographical references and index.
    ISBN 0–465–07649–1
    1. Genetic psychology.   2. Natural selection.   3. Nature and nur-
ture.   4. Neuropsychology.   I. Title.
BF701.G33   1992
155.7   dc20   91–59010
                                                                                CIP

Designed by Ellen Levine

92 93 94 95 CC/HC 9 8 7 6 5 4 3 2 1

In memory of Leon Festinger

Contents

Figures

PREFACE

IN 1989 my old friend Nisson Schechter, a molecular biologist at the State University of New York, Stony Brook, and a man with an enormous appetite for ideas, sent me an article by the immunologist Niels Jerne, written over twenty-five years before. Schechter assured me that the article, in which Jerne threw down a challenge to neurobiologists, was a true classic. He was right. In a crystal-clear and mercifully short essay, Jerne observed that the ideas about selection theory that had been formed in both evolutionary biology and modern immunology must be considered in the light of modern brain science. His insight there leads to an obvious next step, the relevance of selection theory to psychological and social processes. I immediately passed the article on to the smartest man in the world, Leon Festinger, who was then working on a book on medieval history. He had the same reaction, and we discussed the topic for months and decided to have a meeting about it in—why not?—Venice.

Ten outstanding biologists, neuroscientists, and cognitive scientists met to discuss the issue. Manny Scharf laid out its mechanism in immunological terms. Ira Black and Jean Pierre Changeux looked at molecular and cellular aspects of it. Stephen Gould discussed it from an evolutionary viewpoint. David Hubel and Wolf Singer discussed brain development. David Rumelhart looked at the issue in the light of functionalism, while Steven Pinker argued how selection theory could explain basic facts of human language. David Premack talked about comparative psychology development, and Gilbert Harmon considered some of the philosophical aspects of the issues. I fetched espresso and took notes. The conference was the best any of us had ever attended. We all went home to let the ideas simmer.

This book is an outgrowth of that conference—and a perhaps reckless attempt to put together ideas from many fields that may speak to the issue of selection versus instruction. I say "reckless" in the sense that, as a laboratory scientist, I am well aware of the large jumps involved in moving from one level of analysis to another. But a passion of mine, which seems only to intensify with age, is to try and make connections that might explain how the brain enables human cognition. I wouldn't have written this book if I didn't think my ideas are more right than wrong. At the same time, only more work and thought by the hordes of scientists who study these issues will tell the final story.

A particularly significant contribution was made by Steven Pinker of MIT, who is the world's expert on language and on just about everything else in cognitive science these days. It is the argument he made in Venice that I render in the chapter on language. He also read the entire manuscript and provided me with invaluable comments and guidance throughout. I am truly in his debt. Much of the work I cite on brain development has been convincingly reported by Wolf Singer. If there is anything wrong with the argument, it is my fault. The same goes for others whose contributions are duly noted in the text.

The book has also been read by many good friends, including Ira Black, Barry Stein, and Michael Posner, each of whom has given a lot to consider.

I wrote this book at my alma mater, Dartmouth College. While at Dartmouth I taught an undergraduate course on brain mechanisms. What a pleasure it was. There is something captivating about the Dartmouth undergraduate. When I was a student here, the place was all male, and I used to think the mystique was somehow related to the male environment. Not true. The coed version is even better at bringing out the essential feature of these young men and women: They are, it goes without saying, smart. More important, they are feisty and challenging without being snooty. I gave the whole class the problem of selection versus instruction and told them to write about it. Write about it they did, and many of the specific points made in this book grew out of those papers.

I am specifically indebted to Lisa Forlano who not only delivered a clever analysis of mental illness but also helped me analyze the arguments made in Venice. In our laboratory, another student, Robin Plager, researched the work reported on addiction and the effects of the environment on health. We have, visiting from Harvard, Patrick Brown, who assisted greatly on the topic of development and intelligence and also contributed his fine hand in the final editing process. Thanks, too, to Cathy Thomas for her research on psychoanalysis and also for the illustrations. Finally, the careful editing and suggestions of Phoebe Hoss, developmental editor at Basic Books, worked wonders. Jo Ann Miller, senior editor at Basic Books, told me I would need, and love, Ms. Hoss's editorial help. How true.

Selection versus Instruction

F ROM birth to our failing years, the brain is confronted with a seemingly limitless number of messages from the environment. Yet of these countless interactions, how much of what we hear, see, and experience actually influences who we are, how we feel, and who we eventually become? For those of us in the Western world, the answer seems to be that our immediate environment holds the secret to a good, long, and healthy life. We absorb every drop of information about how different substances or dietary practices may affect our brains and our physical health. Americans, especially, are compulsively concerned about which theories to subscribe to, which our children should be raised according to, and which our government should control. The modern parent is forever manipulating the family environment in the firm belief that certain information and products make a difference in the quality of life and mind, and that, somehow, a particular environment is all that is needed to reshape both body and mind to a desirable, utopian end.

Until recently the standard challenge to the importance of the role that the environment plays in our lives has been to suggest that an individual's genetic code is so dominant that there is very little that is open to manipulation by even the most ideal environment. Taken to its extreme, this view proposes that the genes that govern the construction of each person's brain are unrelenting in the specificity they confer on the final

1

nerve nets that support individual behavior. What this means in terms of our ability to help ourselves is that no matter how hard we try to alter our behavior and emotional life through therapy or self-improvement, we cannot succeed because our behavior and emotions are governed by brain organization.

While there is little doubt that there are genetic constraints on any species and that capacities vary within its members, environmentalists have always believed that a large part of the individual variation in a species is due to the organism's being mutable or open to instruction from the environment. They reject the strict nativist view which maintains that variations in the individual can also be accounted for by one's genetic hardwiring. The argument fluctuates back and forth, as environmentalists note that there is obviously instruction from the environment since people learn different languages and different skills. Environmentalists believe that a positive, nourishing environment builds in "good" information and maximizes the capacities of an individual; and that a negative, punishing environment builds in "bad" information, which frustrates and inhibits potential.

Over the years, this simple dichotomy has yielded to the interactionists' view—namely that certain universal aspects of human behavior, such as our capacity for language, are largely determined by genetic processes, while variations in our capacities are shaped by the environment. The data, for example, that show that the capacity for language is derived from genetic capacities is overwhelming. However, few would doubt that experience drives the amount and kind of information we acquire through such a genetically determined system. Thus, one can be a nativist for capacities like language and still believe that individual differences reflect environmental influences. Taken together, today's prevalent views argue for genetic constraints with plenty of leeway built into the brain to allow for modification of behavior through learning. All of this is achieved by a brain that is thought of as malleable, as capable of change by virtue of how its basic elements, the individual neurons, interact and communicate with one another.

In recent years, however, a new challenge to this debate has emerged. Cell biologists, in casting their eyes toward the brain sciences, have forced us to rethink the nature of human mental processes. The issue, referred to as "selection versus instruction," argues in a profound way for a much stronger nativist view. For the selectionist, the absolute truth is that all we do in life is discover what is already built into our brains. While the environment may shape the way in which any given organism develops, it shapes it only as far as preexisting capacities in that organism allow.

Thus, the environment *selects* from the built-in options; it does not modify them.

Few brain scientists, let alone psychologists, have dealt with the implications of this idea, even though, ever since Charles Darwin, the message from biology has been that selection is at work, not instruction. The view that animals adapt to new environments because individual animals modify their body physiology through instruction was replaced with the knowledge that animals that already possess the needed physiology are selected for, and that they will fit into the new environment and survive. The list of examples for selection includes the work of French Nobel laureate Jacques Monod, who, in the mid-1950s, showed that so-called adaptive enzymes are in fact induced by preexisting genes. Also, the Nobel laureates Salvador Luria and Max Delbrück demonstrated in 1943 that bacteria do not adapt and change because of the presence of antibacterial agents; rather, the process is once again selective, and in it, preexisting bacterial alternatives simply blossom. In short, what looks like adaptation in the biological world has always turned out to be a process of selection. Consider the case from immunology.

Immunologists have maintained for years that the body is capable of manufacturing an antibody for any foreign substance introduced into its system. This idea seemed to make sense because of the huge array of substances to which the body can react by making antibodies. It was also known that the body manufactures antibodies to artificial substances that were newly synthesized and had never before existed in nature. It is now known that when a foreign substance assaults the body, a preexisting cell immediately recognizes the intruder and, in order to defend the body, begins to multiply and manufacture proteins. As the cell multiplies, there may be mutations that can, in turn, make proteins even more devastating to the intruder. What this new knowledge reveals is that what once looked like an instruction process (the body *developing* a new molecule in response to the environment) has turned out to be a selection process (the environment *selecting* a preexisting cell the body already possesses to make the appropriate antibody).

It was Niels Jerne who, in 1968, first asked whether this same process could be true for the brain as well. In other words, could most examples of learning be illusionary? Do our decisions and actions result from our discovering what is already in our brains? Is it, as Jerne suggests, that Socrates had it right and poor John Locke had it all wrong? Locke, the consummate instructionist, saw the brain as a piece of blank paper upon which experience is written. He would have held that the mind is unspecified and void of all prior structure. Socrates and the Greek sophists, on the other hand, took a dimmer view of the environment's influence. As Jerne points out, "Socrates concluded that all learning consists of being reminded of what is preexisting in the brain."

In this context, it is important to point out that selection theory describes more than just an aspect of what is innate. One of the mysteries of behavior is its variability. Different animals respond differently to the same environmental challenge. The same animals can respond differently to varying environmental challenges. Selection theory takes this information and points to a new model. The strong form of the argument is that an organism comes delivered to this world with all the world's complexity already built in. In the face of an environmental challenge, the matching process starts, and what the outsider sees as learning is actually the organism searching through its library of circuits and accompanying strategies for ones that will best allow it to respond to the challenge. When this concept, which is well established in basic biology, is applied to more integrative mechanisms of mind it provides more than a clue that the environment and the individual interact; it provides the basic mechanisms by which this interaction takes place.

Establishing such a model is important because the prevalent assumption of modern brain science, that instruction processes are at work, has yielded precious little insight into the brain mechanisms involved in memory and learning. Selection theory provides a link by which knowledge of how genes and environment interact can be bootstrapped to issues of cognition. In this link, the case from immunology is crucial. There the selectee (the antibodies) are sophisticated and complex from their beginnings, having all the rich information. Yet, as is well known from Darwinian theory, organisms mutate slowly, albeit surely, over millions of years. Only slowly do they build up changes in neural-circuit function. In this sense, the complexity is in the selector (nature), not in the selectee (the individual organism). Thus, a selection system could be richer and more flexible than an instruction system. Allowing millions of years of evolution to deposit in our brains circuits that allow us to adapt to complex environmental challenges might be our savior as a species. Surely upon examination, we will discover that the complexity is not in what the selectee knows, but in what the selector has done over millions of years of evolution.

If this hypothesis is accurate, it is quite possible that we humans are living in a delirious frame of mind about what influences what and what we can do about it. The deceptively simple notion of applying biological constructs to psychological processes challenges our whole philosophy of life—including the importance we place on personal achievement, intelligence, and acquired beliefs. Even though at the psychological level much of what happens to a person *appears* to be the result of instruction, at the molecular level we consistently see signs that *selection* is operating. Taken another way, the hypothesis implies that a complex organism (the human) functions adequately in a complex environment, at least at the molecular level, and knows nothing about that environment except that it is foreign. The organism knows only its own reactions to the quality of foreignness. Yet while the hypothesis appears to be true at the molecular level, must it also be true for the next level of organization? In other words, is there any way in which molecules that satisfy the selection hypothesis at the molecular level are strung together to make up neurons and, eventually, neural circuits that are responsive to instruction? While every brain scientist assumes this to be possible, is it really the case? At higher levels of psychological processes can one truly distinguish between what looks like instruction, such as learning calculus, and selection processes working at more molecular levels? If, indeed, selection theory does operate at the higher level of "whole-brain" processes, we must seriously rethink our current conception about the nature of psychological processes.

It is my aim to show that the selection process governs not merely low-level neural circuit events like synaptic relationships (or how neurons talk to each other), but also the complex circuits responsible for higher functions, such as language and problem solving, and that, indeed, these were built into the brain as the result of millions of years of evolution. In pursuing the larger issue, I will draw on the work of dozens of investigators, both classic and contemporary. As Michael Polanyi argued in the 1950s, in order to understand anything, whether it be a frog or the human mind, one has to define its function, and then examine it from an evolutionary perspective. To do this—as the young social scientists John Tooby and Leda Cosmides have argued—is to reject simplistic ideas about the evolution of behavior and assume that evolution produces *cognitive* mechanisms; and that these mechanisms produce and drive the multiple and varied behavioral responses we humans generate in response to environmental challenge. Thus, it is not behavior that evolves, but cognitive mechanisms, such as our capacity for language, or problem solving, or categorization, and so on.

This fundamental point allows evolutionary psychologists to predict wide variations in human behavior. With each new environment, a cognitively driven organism would allow particular behaviors befitting the local challenge. Further, since all members of our species are equally adaptable across a variety of challenges, it becomes clear that specialized systems have evolved at the cognitive level. In short, with Tooby's and Cosmides' view comes the possibility of directly comparing the basic results of biology with those of higher psychological processes. Both body and mind come with complex structures from which the environment selects a response.

In contrast to many current views of the human mind that maintain that human evolution has produced a general-purpose problem-solving device with a more or less infinite capacity for problem solving, I believe, along with Tooby and Cosmides, that a group of species-specific devices, such as language systems, are now in place in us as a consequence of evolutionary events occurring thousands of years ago in the Pleistocene era. If modern cognitive neuroscience knows anything, it knows that the human brain is replete with specific processing systems. Tooby and Cosmides provide a possible framework for our understanding of how this came to be.

For years, mind scientists have grappled with the problem of how much information an organism must have before it begins to deal with its environment. In the late 1950s, when Noam Chomsky proposed that language was largely made possible by a special device (organ) built into the brain, psychologists went berserk. B. F. Skinner, the ultimate environmentalist, attacked Chomsky vehemently, but to no avail. Even today, there are still those who still try to demonstrate that environment dictates all. The current fad of cultural diversity is the most recent form of the environmentalist argument, wherein we are told that nothing exists except as it does in interaction with the environment. Here it is proposed that no idea or mental representation is biologically real, constrained, or more important than any another, since we are all products of the environment in which we live.

Yet while scientists continue to stumble over themselves, science progresses. One of the current realities emergent from the work in artificial intelligence (AI) is that a machine starting from scratch, just like a Lockean baby, cannot learn complex things like language. The AI people complain about what they call the "frame" problem. How do you set the boundary conditions of what is to be learned? There needs to be an initial plan (or "architecture," as it is now called) that sets the stage for what is to be learned. Without it, no artificial device, such as a computer, knows what to do. Having these special-purpose devices built into the brain solves the problem of the human species.

Modern humans have been playing in a very long ball game whose rules are complex and frequently appear at odds with our experience. Too much of academic psychology and modern brain science have been looking at and inspecting only a small aspect of the game—say second base. They take it apart, study it, theorize about it and all the rest, but fail to see that second base is part of a much larger and more dynamic set of circumstances. In this book, my objective is to present the larger biologic and evolutionary context for considering the human mind. It is my hope that you shall discover, as I have, that all the ways that human societies try to change minds and to change how we humans truly interact with the environment are doomed to fail. Indeed, societies fail when they preach at their populations. They tend to succeed when they allow each individual to discover what millions of years of evolution have already bestowed upon mind and body.

CHAPTER I

A Lesson from Biology: Modern Immunology

WHEN a puff of air is presented to the eye, we immediately blink. When our knee cap is hit just right, our leg jerks. Such events are examples of innateness—a genetically powered, built-in, and immediate response to an environmental challenge. There are thousands of commonplace examples of such immediate responses, and they are well understood. It is not so clear that the delayed responses, such as learning a language, problem solving, or creating mental images also arise from genetically constructed prewired neural circuits that have been built in by natural selection over millions of years. Yet, there is a growing body of evidence that these higher order processes are also built in. They are referred to as domain specific Darwinian algorithms—which is a fancy way of saying evolutionary pressure has coughed up specialized circuits in the brain that carry out specific mental functions. When we think we are learning something, we are only discovering what already has been built in to our brains.

Astonishingly, this kind of built-in system is known to exist at the cellular and molecular level—in the immune system. Great complexity is built into the immune system, and selection works on that complexity. There is no quibbling about this: That is the way the immune system works. And, as we shall see, selection processes are also central to evolutionary theory and all of biology. There is a great deal at stake if this claim is true for human cognition. It casts a different light on the nature/nurture argument that has plagued science for years, and makes obsolete traditional and popular views of psychological processes, such as those expressed by every behaviorist and many knowledgeable observers of molecular biology, such as Masimo Piatelli-Palmarini. As the cognitive scientists Steven Pinker and Paul Bloom note in their already classic paper on the evolution of language, it was only a few years ago that Piatelli-Palmarini claimed, "since language and cognition probably represent the most salient and the most novel biological traits of our species . . . it is now important to show that they may well have arisen from totally extra-adaptive mechanisms." In other words, language and cognition are acquired skills that come about by learning. It is one thing to talk about built-in antibodies and quite another to talk about built-in knowledge systems.

In order to get to the bottom of these ideas and to see if the model from immunology can truly apply to mechanisms of mind, we must grasp how profoundly and pervasively selection processes are known to work within each and every one of us. Our immune system is the model for these ideas, and it is fascinating to see how it responds to outside challenges.

Modern Immunology and Selection Theory

There are, in the world, millions upon millions of biological structures, such as organisms and chemicals. Some of them are in our body, but most are not. When those that are not in our body do enter it, our immune system generates a massive response to rid the body of these foreign invaders. When I was in graduate school in the early 1960s, the leading biologists and chemists thought that the immune response was accomplished by instruction mechanisms, and even likened the process to learning: That is, the antigen (or invader) instructed certain cells to form antibodies, which in turn protected the organism from damage. The idea grew out of the reasonable assumption that a body can make more antibodies than could have already been specified by its genes. Absolutely any natural substance, when injected into the body produces a specific antibody. Additionally, every species has its own species-specific antigens, and each of these can produce a specific antibody in another animal. The instruction idea seemed entirely reasonable, and mechanisms like these were proposed to explain how it might all work. An antibody is a specific

9

molecule, called a "protein." When the body is challenged, the protein molecule folds in and around parts of the invader and eventually locks it up and neutralizes it. Linus Pauling, the Nobel laureate and chemist, was one of the scientists to whom this idea made perfect sense.

Another compelling reason for this view of antibodies was the discovery that molecules made by humans, which had never before existed in the natural world, can also trigger antibody formation. Thus, an artificial organic compound like dinitrobenzene, when injected into rabbits, caused them to develop antibodies to the synthetic agent. After analyzing this type of data, scientists concluded that the body's immune system recognizes this oddball invader and forms a response to it. Specifically, as with natural substances, it was thought that the giant protein molecules, the δ-globulin, float around within the body in an unfolded state; when the foreign element (antigen) is introduced, it folds around the molecule, and the shape is locked into place by the formation of other chemical linkages (fig. 1.1). The antigen is then released and the new instruction-sensitive protein (antibody) has on it an active site that can respond to other antigens of the same kind. Thus, the first antigen would serve as a template for the unfolded protein molecule and continually roam the body preparing antibodies for itself.

The instructionist view was based on the specificity of the antibody response. For example, since changing one small element of a long chemical chain had been shown to produce a different antigen, it followed that one amino acid that varied out of the hundreds that make up the human immunoglobulin molecule could induce that molecule to trigger the production of a different antibody. By the mid-1960s, scientists were beginning to wonder why—if the immune system is exquisitely sensitive to slight variations in molecules—the body does not trigger the antibody response to the millions of antigenic agents within itself. If virtually any molecule or cell from animal A will produce an antibody response in animal B, how does the body know to challenge only foreign antigens and not its own? It was reasoned that a particular body could not first compare a foreign antigen to a library of its own millions of antigens because the process would simply be too time-consuming to be completed efficiently. The bold conclusion was, therefore, that each animal is born with all the antibodies it will ever need. They go to work immediately on any challenge. From this insight, Darwin's idea of selection appeared in the context of immunology. Antibodies are chosen from a *preexisting* library of antibodies in an animal. In short, there is no instruction.

It should be noted here that selection process, in the traditional Dar-

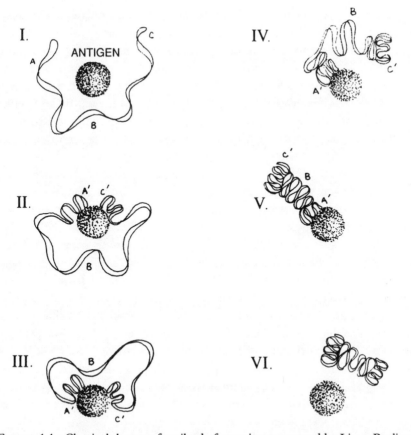

FIGURE 1.1   Classical theory of antibody formation promoted by Linus Pauling. The six-step process follows from an antigen's initial interference in the folding of the globulin molecule. Steps I–IV depict the folding process. Step V represents the fully folded molecule, and Step VI shows the antibody free to go on the attack. Based on figure 14-1 in J. E. Cushing and D. H. Campbell, *Principles of Immunology* (New York: McGraw-Hill, 1957).

winian sense, means something different. Frogs do not wait around for millions of years to be selected and multiply. Frogs are now a homogeneous group. They vary in their makeup, and should a change occur in their niche, some of them may be able to adapt to it because of a preexisting capacity—a capacity they have but that other frogs may lack. Those that have the capacity survive, and those that do not, die off. There is selection from a group of a preexisting capacity but in a different sense from that present in the immune system. In the immune system antibodies are selected, but those not chosen after a particular challenge remain in readiness for the next challenge.

11

Evidence for selection mechanisms began to accumulate in immunology as it was noted that cells containing intracellular antibodies do not have any antigens attached to them. Even with the variety of new chemical tricks available, it was impossible to find any suggestion that these antibody-producing cells were triggered by an antigen! Such data contradicted the idea that an antigen serves as a template for antibody production. In other words, the invading antigen had to be different from the ones already existing in the host organism. If it wasn't different, it would not be considered "foreign" and no antibody response would occur. By the time some matching or comparison process took place, whereby the invading antigen was assessed, days, weeks, or possibly months could go by. By that time, the foreign antigen might have already done its deed and destroyed the organism.

In other experiments, antibodies were treated with other chemicals that eliminated their folded structure. Specifically, an antibody that was made to fit something like guanidine was unfolded, and the guanidine removed through a washing procedure, leaving only the antibody (fig. 1.2). Yet, these new unfolded antibodies were shown to be likely to bind to new challenges of guanidine. This finding suggested that, in its primary structure, the molecule possesses enough information to allow for a spontaneous refolding after the tearing down of its structure and its washing, and thus, can once again serve as an antibody to the appropriate antigen. The idea that the antibody was formed as the result of instructional processes appeared less and less likely.

The implications of this theory, as first pointed out by Niels Jerne (fig. 1.3), are enormous for a theory of biological and psychological systems; and we must understand in more detail both the mechanism by which antibodies are produced and their chemical structure. Following the mechanisms of cellular action is like playing chess: There is a large board with players, many of which have both local and far-reaching effects on the other players. Like all hard sciences, there are real facts with which to contend, and real mechanisms of action to be developed.

Antibodies are produced by special cells called "lymphocytes." Depending on species and age, animals have between 10^9 and 10^{12} different lymphocytes: That is, animals possess at least 1 billion different lymphocytes, and some have as many as a 100 billion lymphocytes. When properly stimulated, each of these cells has the genetic information to make a specific antibody. This fact alone suggests that a selection process is responsible for antibody production. Surely the billions of antibody-producing cells in the human could match and respond to any antigenic structure.

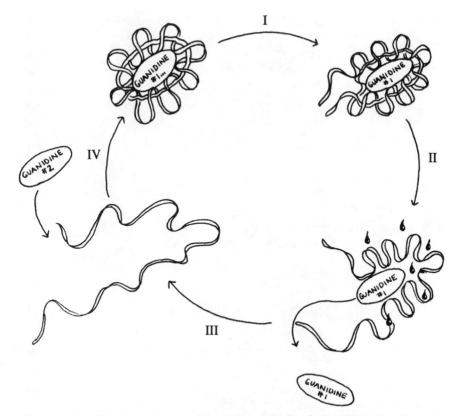

FIGURE 1.2 The classical view was challenged by a finding like this. An antibody, once formed, can be torn apart only to be reformed without the stimulating antigen. The fact that it can still function as an antibody suggests that the original antigen did not impose its shape.

A lymphocyte produces antibodies that stay lodged on the surface of its cell membrane. When a foreign antigen comes along, it somehow finds the appropriate cell(s), and combines with the lymphocyte cell surface. This reaction triggers the cell to multiply and divide and thereby produce more cells. These "daughter cells" produce more free-floating antibodies, in addition to the antibodies already lodged in the cell's own surface. These free-floating antibodies roam the body looking for more of the antigen to annihilate.

Although there is some truth in the old theory, little is actually known about the particular mechanism of cell-to-cell recognition. At one time, the antigen-antibody response was likened to a "lock and key" relationship. This simile has fallen out of favor, however, as it has become apparent that the human body has some 30 to 40 billion lymphocytes, and

FIGURE 1.3   Niels Jerne, the Nobel laureate immunologist who first drew attention to the immunology model as applying to other biological and psychological systems. Source: James T. Barrett, *Textbook of Immunology: An Introduction to Immunochemistry and Immunobiology* 5th ed. (St. Louis: C. V. Mosby Co., 1988).

any one antigen might consist of several million molecules. We know neither how they find each other, nor how much time is spent in molecule-to-molecule exploration of each other to find the right site for contact. Since a low dose of antigens has been found not to produce a detectable antibody response, we know that the body does not muster an antibody response to every foreign invader.

In order to understand some of the loopholes in the system, other concepts need to be explained. For instance, antigens and antibodies combine with different affinities. There is high-affinity binding, low-affinity binding, and everything between. As you might guess, low-affinity binding is not effective, as the antibody has a tough time neutralizing the antigen. The body responds to this state of affairs in a fascinating way. Sufficiently strong binding between antigen and antibody stimulates the lymphocyte to divide and to produce more antibody. This process is going on in millions of cells at a time, making inevitable some mutations. These mutations (which now possess a new genetic combination) will produce antibodies with a slightly different structure, which may yield a better fit for the original antigen and bind to it. This cell, in turn, starts to divide, making possible still further mutation and a better fit. As the overall process goes forward, the original antigen is dealt with more and more efficiently. The vast number of viruses and other antigens to which organisms are exposed indicates how crucial the mutational mechanism is to the largely successful manner in which our bodies ward off challenge.

A unique event that assists the immune response has recently been discovered. In this process of "hypermutation," the normal rate of mutation increases, making more probable the creation of a lymphocyte with a higher affinity for the challenging antigen. While the details of this process are not yet understood, Manny Scharff and his colleagues at the Albert Einstein School of Medicine have demonstrated that the mutations occur on one part of the antibody molecule (fig. 1.4).

There is nowhere in all this, however, any instruction from the environment. The antigen simply lies in wait and selects out the most appropriate antibody molecule, which either preexisted in the body or evolved from strict genetically driven processes. This overwhelming fact is consonant with much of what is known about every other biological system— from cellular molecular response to the evolution of the species.

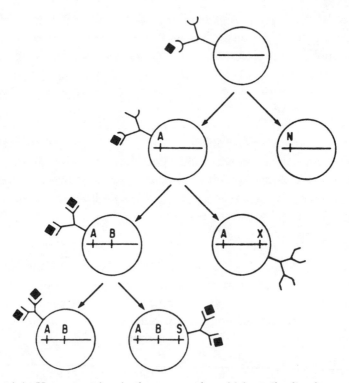

FIGURE 1.4 Hypermutation is the process by which antibodies become ever attuned to a challenging antigen. Figure 1.4 shows a tree diagram of this process. The antibody at the top has a binding site with low affinity for the antigen: that is, the antigen does not fit well into the binding site. However, once antigenic stimulation and hypermutation occur, a change in the cell's gene sequence by a replacement (A) or a nonsense (N) mutation occurs. Those cells with the replacement mutation undergo changes in structure so as to increase their binding affinity for the antigen. The nonsense mutation causes cells to cease proliferation. The process continues, more mutations are introduced (B and X), and the affinity for binding improves to the point where a precise fit between antigen and antibody has been achieved. Silent mutations (S) that do not result in amino acid substitution can occur at any time. Source: Deborah L. French, Reuven Laskov, and Matthew D. Scharff, "The Role of Hypermutation in the Generation of Antibody Diversity," *Science* 244 (4909 [1989]): 1152–57. Reprinted with permission.

EVOLUTION AND SELECTION THEORY

The immunologic evidence firmly establishes that selection is a strong force in biological processes. It is the key idea in Charles Darwin's theory of natural selection and, with the help of hindsight, the general concept of selection seems almost obvious. But as Stephen J. Gould has pointed out, it wasn't always that clear. In order to understand the brilliance of Darwin's discovery, and now its broad application to biological sciences, we have to take a step back into the early nineteenth century and consider the classic argument of structuralism versus functionalism.

The structural/functional dichotomy of nineteenth-century biology centered around whether form precedes function, or function precedes form. The structuralists maintained that physical forms have set capacities and are able to carry out specific functions. If those functions are desirable in a particular niche, the organism survives; if not, the organism dies. The argument seems wonderfully simple, logical, and true.

The functionalists, on the other hand, maintained that a function is needed to survive in a particular niche, which shapes the structure of the organism to supply the much-needed function. This was the position of the French naturalist, Lamarck. While what he actually said has been somewhat overdrawn by modern biologists, he argued, as Gould states, "that it is the function of the organism responding to changed circumstances that sets its form." Thus, Lamarck—though every bit as much a functionalist as the great Charles Darwin—simply put his cards on the idea that the environment instructs the organism to change.

In 1809 Lamarck argued for an evolutionary theory with two separate processes (fig. 1.5): the first, a force in nature that "tends incessantly to complicate organization"; the second, the "influence of circumstance," which he felt produced lateral adaptation forces. These two ideas produced two principles of heredity for Lamarck: the inheritance of acquired characteristics and the principle of use and disuse. Lamarck thought that the environment changes slowly; and that, during its life, the organism changes to meet those demands. When the organism dies, it passes on to its offspring this gained and hard won knowledge. When an organism does not use particular capacities, they disappear as characteristics of the species. Gould has likened Lamarck's ideas to the Protestant ethic, and somewhat whimsically suggests that this is the reason they may have been so readily adopted. If you work hard and develop something, you keep it. If you don't use it, you lose it.

What is interesting about Lamarck's ideas is how close they come to explaining the phenomenon of adaptation. After all, animals do "adapt" to new niches. The ones that do, survive. Given those two facts and in the absence of modern biological knowledge, Lamarck's explanation was indeed plausible.

But it took Charles Darwin to get things right. He first articulated the theory of natural selection as a mechanism explaining not only evolution,

FIGURE 1.5 Charles Darwin *(left)* and Jean de Lamarck *(right)* were remarkably close in their ideas. In modern terms, Lamarck was an instructionist, and Darwin, a selectionist. Sources: (Darwin): The National Portrait Gallery, London; (Lamarck): Archives du Musée d'Histoire Naturelle, Paris.

but also, to some extent, speciation. The notion that animals adapt to their environment, and change their strategies for living in order to survive, belies the underlying fact that animals are selected out by an environment. Animals that are genetically equipped to handle the environment survive, and those that do not have that necessary equipment die. In short, Darwin also had two main principles concerning evolution: The raw material of an organism has been established by genetic mechanisms, and the environment gives direction to that raw material by selecting out the best of it. These two processes must be absolutely separate, and modern research has shown them to be. Yet the important difference, of course, is that the variation in raw material is not directed toward adaptation. It exists in a population as a consequence of a fundamental biological process. There is no intelligence to the variation. Over evolutionary time, however, complexity accumulates in the organism as it comes to adapt to more and more new challenges. At the level of immunology and antibodies, the complexity is already there prior to the selection process working. In evolution there are thousands of examples of this process.

Consider a recent example that occurred on Darwin's beloved Galápagos Islands. In 1977, a major drought caused a dearth of small seeds that are the primary food supply of the Darwinian finches. The finches normally crack open the small seeds with their beaks (fig. 1.6). After the drought, there were only large seeds, which, though not affected by the drought, were too big for the finches to eat. As a consequence, the population of Darwin finches dwindled—to be gradually replaced with another group of finches that had larger beaks, beaks that could crack the large seeds. What happened? There was adaptation by a shift in the population. Existing in the finch population all along, though few in number, were birds with larger beaks. As a result of the new environmental challenge, the big-beaked birds survived and multiplied. Thus, no individual bird changed its composition. A population shift made it appear that the finches had acquired new characteristics.

The same sort of mechanism is responsible for a species' dropping useless characteristics. As Renato Dulbecco, a virologist of the Salk Institute, has pointed out in his fascinating book, *The Design of Life*, if flying birds are introduced to an island where there are no predators, the

FIGURE 1.6   The Galápagos Islands. The drought of 1977 presented an environmental challenge to the preexisting population of Darwin finches causing a population shift to those finches fit to survive in the "new" environment.

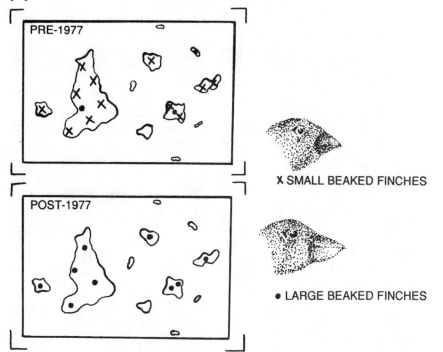

X SMALL BEAKED FINCHES

• LARGE BEAKED FINCHES

19

ability to fly may be harmful to the species. Flying requires energy, and the bird that is flying around doesn't have much energy for acquiring food. Ultimately, the bird that continues to fly will lose out to the variant birds of the species that can't fly and busy themselves reproducing. Thus, selection forces help rid a species of useless characteristics. Darwin recounted similar examples, observing that insects evolving on small windy islands lost their wings so as not to be blown out to sea. Such examples have persuaded most people that characteristics of the species have slowly evolved through geological time.

Some recent thinkers have argued, however, for what is called "punctuated equilibrium"—an evolutionary process whereby speciation happens by an explosion of mutational events followed by years of stasis. The idea was first proposed by Stephen J. Gould and his colleague, Nigel Eldredge, and has been expanded on by Gould over the past fifteen years (fig. 1.7). To support this notion, in his irrepressible manner, Gould has even tried to resurrect the theory of a discredited biologist, Richard Goldschmidt, and argue that there may be reason to believe in something similar to, but not quite, the Goldschmidtian idea of "hopeful monsters."

FIGURE 1.7   Stephen J. Gould *(left)* and Richard Dawkins *(right)* have both made enormous contributions to the public understanding of evolutionary process. Sources: (Gould) Courtesy of the Harvard University Press Office; (Dawkins) By permission of Richard Dawkins. Photo © by Lisa Lloyd.

Goldschmidt believed in "saltation." He believed evolution occurred in big jumps like, say, two *Homo erectus* parents giving birth to a *Homo sapiens*. This idea holds that a sudden mutation event might produce a whole new entity, such as a monster of some kind. Metaphorically, Gould argues that punctuated equilibrium may produce new species with lightning speed, although Gould's punctuated equilibrium works in geological time, which involves units of a hundred thousand years. More relevant to our concerns, punctuated equilibrium has been put forth as a process by which cognitive activities such as human language might suddenly explode on the screen. Steven Pinker and Paul Bloom have pointed out that the idea of punctuated equilibrium is not excessively radical, and surely does not reject the idea that changes in speciation usually occur over long periods of time, but that there seem to be bursts of change as well. After all, we know there have been long periods of stasis in evolution when nothing appeared to have been happening, when after millions of years a species is seemingly the same. Suddenly the geological record shows a change, and a new species is on the scene. The current heated debate is over what may cause that apparent sudden change.

Richard Dawkins, the British zoologist and author, argues that existing data from the fossil evidence are too ambiguous to allow for anything but the interpretation that most evolution occurs gradually. Wherever a new species suddenly appears in the record, Dawkins points out that it might be an artifact of fossil sampling. His view is that big evolution often occurs in isolated populations where genes are not diluted. Then, once a change has occurred, the new organism invades the general population, takes over, and eventually becomes part of the large population's fossil record. It is only to the geologist looking back in time at his record that a sudden event appears to have occurred.

While all of this may sound like splitting hairs, and a little off our mark, it bears on how complex psychological functions are acquired through selection mechanisms in the human species. We have to unearth a viable mechanism for the manner in which complexity gets built into the human brain. Traditional evolutionary theory wants it to have developed there gradually, whereas the punctuated equilibrium proponents see complexity as occurring over a shorter time scale and all at once. Yet it is truly difficult to imagine how serious complexity could suddenly be built into an organism. One organ over which the gladiators of evolutionary theory battle is the eye.

Along with Pinker, Bloom, and others, Dawkins pointed out that, because of its inherent complexity, there is no way an eye could suddenly appear in an organism (fig. 1.8). Coordinated changes would have to occur in the lens, the muscles that control the lens, the retina, the connections of the retina to the brain and so on. Since each aspect of the visual system is under different genetic control, orchestrated changes in all of these systems are out of the question. The eye has to have evolved gradually. To this, Gould said, "We avoid the excellent question, What good is 5 percent of an eye? by arguing that the possessor of such an incipient structure did not use it for sight." Of course, as Pinker and Bloom have pointed out, "no ancestor to humans literally had 5 percent of a human eye; the expression refers to an eye that has 5 percent of the complexity of a modern eye." Then, in 1986, Dawkins responded: "An ancient animal with 5 per cent of any eye might indeed have used it for something other than sight, but it seems to me at least as likely that it used it for 5 per cent vision. . . . Vision that is 5 per cent as good as yours or mine is very much worth having in comparison with no vision at all. So is 1 per cent vision better than total blindness. And 6 per cent is better than 5, 7 per cent better than 6, and so on up the gradual, continuous series."

To my mind, it is clear that punctuated equilibrium is not the evolutionary mechanism underlying the appearance of cognitive advances such as language. Like the eye, language is too complex to have suddenly appeared. The underlying issue here is that punctuated equilibrium, like the spandrel theory reviewed below, is a proposal for how evolution proceeds that, at its core, is more consistent with how an empiricist would want to view evolution. The authors of such ideas would have us believe that the human being is an accident of such mechanisms, not a slowly evolving and adapting creature shaped by millions of years of environmental challenge.

It is worth noting in context of this ongoing debate that Gould is considered by some to be singlehandedly responsible for reintroducing discussions on evolutionary theory to the general public. His spellbinding articles and his formidable lecture style inspire and challenge all. At a small conference in Venice that I organized in 1988, Gould took some of the participants on a private tour of the basilica of San Marco, spelling out to us his now famous and intriguing "spandrel theory" and its relation to problems in brain evolution. As we walked around the cathedral listening to this immensely talented man lecture not only on evolutionary theory but also on art and history, a small crowd began to gather for this remarkable event. There was even a priest or two!

Gould was warning the anatomists that the structures they were studying and trying to understand may have evolved for other purposes, and

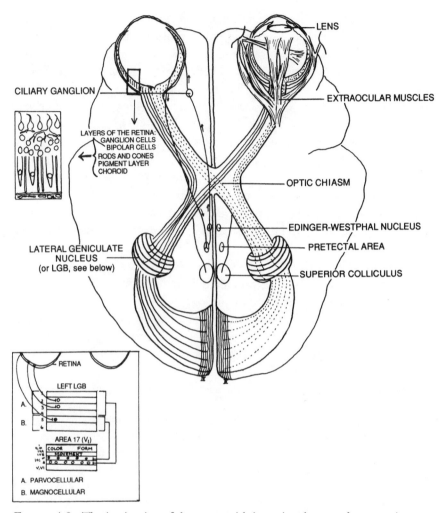

FIGURE 1.8 The intricacies of the eye—with its retina, lens, and connections to specific points in the brain—make it unlikely that such structures can appear spontaneously. Adapted from Arthur C. Guyton, *Basic Neuroscience: Anatomy and Physiology* (Philadelphia: W. B. Saunders, 1987), p. 304; H. B. Barlow and J. D. Mollon, eds., *The Senses* Vol. 3 *Cambridge Texts in the Physiological Sciences* (Cambridge: Cambridge University Press, 1982), figures 2.1 and 2.5 (facing pp. 35 and 43); and Bryan Kolb and Ian Q. Winshaw, *Fundamentals of Human Neuropsychology* 3d ed. (New York: W. H. Freeman, 1990), p. 213. Copyright © 1990 by W. H. Freeman and Company. Reprinted with permission.

may only recently be performing the function the modern brain scientist is trying to understand. It is this sort of observation and reasonable possibility that sends chills down the spine of brain scientists. Could one be nonsensically looking for the antecedents of a brain structure that, in a primitive animal, may have been doing something else? All of this stems from his idea on spandrels, an idea that was triggered by an earlier visit by Gould to San Marco (fig. 1.9).

At the core, Gould's idea is non-Darwinian. He is suggesting that evolutionary changes may occur in ways that do not involve selection. When this notion is taken together with his other ideas on punctuated equilibrium, he appears, to some popularizers of science and to the casual reader, as someone who does not believe in natural selection. Nothing could be further from the truth. Even though he is unquestionably trying to augment strict Darwinian principles of natural selection, he is fundamentally as much a Darwinian as anyone else. After all, natural selection could be taking place during punctuated equilibrium. Still, if readers of this book know anything about evolutionary theory, they will most likely know about Gould's ideas and the color he has given Darwinian theory. Consider his definition of spandrels and Pinker and Bloom's subsequent interpretation of his claim:

> Spandrels: the tapering triangular spaces formed by the intersection of two rounded arches at right angles . . . are necessary architectural by-products of mounting a dome on rounded arches. Each spandrel contains a design admirably fitted into its tapering space. An evangelist sits in the upper part flanked by the heavenly cities. Below a man representing one of the four biblical rivers . . . pours water from a pitcher in the narrowing space below his feet.
>
> The design is so elaborate, harmonious, and purposeful that we are tempted to view it as the starting point of any analysis, as the cause in some sense of the surrounding architecture. But this would invert the proper path of analysis. The system begins with an architectural constraint: the necessary four spandrels and their tapering triangular form. They provide a space in which the mosaicists worked; they set the quadripartite symmetry of the dome above.
>
> Such architectural constraints abound, and we find them easy to understand because we do not impose our biological biases upon them. . . . Anyone who tried to argue that the structure (spandrels) exists because of (the design laid upon them) would be inviting the same ridicule that Voltaire heaped on Dr. Pangloss: "Things cannot be other than they are. . . . Everything is made for the best purpose. Our noses were made to carry spectacles, so we have spectacles. Legs were clearly intended for breeches, and we wear them." Yet evolutionary biologists, in their tendency to focus exclusively on immediate

FIGURE 1.9 The Basilica of San Marco and its now famous spandrels. Shown here, a seated evangelist, and below, personification of river. Source: Stephen J. Gould and R. C. Lewontin, "The Spandrels of San Marco and the Panglossian Paradigm: A Critique of the Adaptionist Programme," *Proceedings of the Royal Society of London* B205 (1979): 148.

adaptation to local conditions, do tend to ignore architectural constraints and perform just such an inversion of explanations.

(Gould and Lewontin, 1979, pp. 281–88)

Gould wants to use this idea to paint a larger canvas upon which evolution can work. Gould feels the nonadaptationist views of evolution have been unfairly maligned. He is certainly reverting to some of the excessive tenets of sociobiologists. Gould wants caveats in traditional Darwinian theory, and this is the case he more or less advocates for this aspect of evolution. He doesn't want to be part of a species that finds all of its traits derived from natural selection. He wants culture to take some responsibility too.

Yet Gould has a problem. Complex structures simply cannot emerge by any means other than through natural selection. As Dawkins, Pinker, and Bloom have shown, it would be absurd to argue that the eye or language or domain-specific algorithms happened to be the result (spandrel) of the evolutionary changes going on elsewhere in the body. It is a nice try on behalf of the Lockeans, but it doesn't work.

EVOLUTION AND MODERN MOLECULAR BIOLOGY

Having said all of this, there are indeed periods of stasis in evolution followed by new useful structures. We need to understand the manner in which biological processes are continually working at the level of the genome with those working at the level of the whole organism. We need to know how molecular mechanisms explain both the apparent changes and the formation of new species on some time scale.

It has been known for decades that animals can exist comfortably for thousands, if not millions, of years with a fairly defined set of characteristics. Then they are presented with a change in the environment. In human evolution, for example, the quality of tool use did not change for about a million years. Suddenly, tools were markedly refined to a different and better quality. That change appeared too quickly to be attributed to mutational forces. When animals gradually change, the changes can be attributed to small mutations working in harmony with selective forces. Yet large and quick changes are more difficult to understand. Are large-scale changes evidence for the view that the environment can induce changes in the organism?

In answering this dilemma, Renato Dulbecco points out that it is just as difficult to explain the long periods of stasis in the history of a species. After all, since mutations are always occurring and cannot be prevented, why are no changes seen for these long periods of time? One possibility is that the changes do in fact occur, but lead to neutral consequences for the animal.

The genetic mechanisms active in a species are active in both an internal and an external environment. In the internal environment, a complex set of constraints works for and against any major mutations. During the development of an organism, a complex set of interactions works throughout the body to allow for a successful birth. There are thousands, if not millions, of interactions between chemical and cellular elements, each of which is generated by hundreds to thousands of different genes. If a mutation changes the composition of one of the elements or the timing of these events in any way, an abnormal entity could result. In that event, the organism would most likely self-destruct.

Another, more subtle, method of change comes about very gradually. Such neutral change occurs when certain genes become different in their action, but do so in a constant environment, so that the difference is not noted. Should the environment change, however, these genes have prepared themselves to click in and show their usefulness. This kind of change in the genome, called a "genetic drift," is a powerful force in evolutionary change. One example of genetic drift, which is to say how a potential change in genes waits for exposure, occurs in genes that control the histocompatability process—a process by which the immune system can differentially attack foreign cells while simultaneously sparing genetically compatible cells. A gene that would normally control the response to a particular virus might have undergone a change that remains hidden so long as that virus does not appear in the environment. If it should appear, the animal could be seriously affected due to the hidden change, and the animal could die.

These microchanges in the genetic posture of a species are tightly controlled by its internal environment. Genetic drifts do occur and set the stage for the big changes, which can occur only when there are big changes in the external environment. Such a big change is believed to have occurred in the Turkana, where a calamity in the environment caused dozens of species to die and others to emerge. Gould recounts a similar event which occurred in the Burgess Shale in the Canadian Rockies. This event is especially interesting for its illustration of the wonderful luck that found our species existing at all.

In our current world, the gene for thalassemia (a condition of abnormal hemoglobin formation that produces anemia) also enables the body to ward off malaria. Where malaria is prevalent in the world, over 20 percent of the population carries this gene. If malaria should spread throughout the world, so would the gene for thalassemia (fig. 1.10). Again, the microevents of the body are set into a dynamic equilibrium with the real environment.

The evolutionary history of our species, together with the nature of the immunologic system demonstrate the powerful force of the selection process. In every instance where the environment was thought to be

FIGURE 1.10 The gene for thalassemia will spread throughout the world in the wake of malaria.

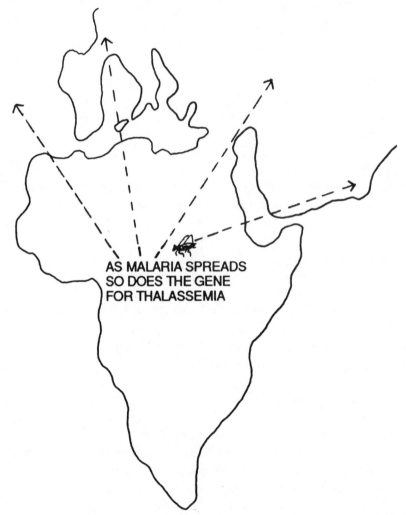

AS MALARIA SPREADS
SO DOES THE GENE
FOR THALASSEMIA

instructing the organism, a feature of the organism or the organism itself was being selected out. With a system as locked in as this, the question is, how much information do we in fact accept from the environment?

In some sense, the underlying notion I am exploring in this book, that the immunologic and evolutionary model of selection can explain both our brain and our behavior appears almost nonsensical. We commonly think that we are learning new things: A French child learns to speak French, a Japanese child speaks Japanese. How does the idea of selection even begin to apply to these cases? The fact is that some truly dumb animals are served by a biological system that is carrying out the most complex computations of space, time, and energy. Most of these activities go on outside of the awareness of animals and humans. Indeed, they go on automatically and are controlled by peripheral processes that have been put in place by selective pressures over millions and millions of years—like the smart computer chip upon which a set of instructions to carry out a specific task is hard-wired and set.

In this light, acquiring the capacity to speak French or English—or, for that matter, any capacity—may reflect little more than a specific environment guiding one of dozens of built-in systems to arrange and process information in a way relevant to that environment. Learning may be nothing more than the time needed for an organism to sort out its built-in systems in order to accomplish these goals. Of course, this is a generous interpretation of how selection mechanisms affect our psychological lives. Deciding whether this line of thinking is plausible requires first a consideration of the nature of the brain itself.

CHAPTER 2

The Plastic Brain and Selection Theory

ALTHOUGH there is little argument that the selection process is at work at both the molecular and the evolutionary levels in whole organisms, there are major questions about whether the brain develops and functions in accordance with the concept of selection. Now if, as I argue in this book, the majority of our psychological capacities are the result of natural selection, the developing and static adult brain, which houses the neural circuits that enable the human's high-powered psychological mechanisms to exist, must develop in a surefire, genetically determined way. At the level of behavior, for example, we want to see whether or not a baby *learns* to identify a face, or whether there exist in the brain specific circuits enabling facial recognition, circuits laid down by prior genetic forces arrived at through selection pressure.

In considering how complex perceptual functions such as facial recognition—or the even more complex functions such as language or simple learning of anything—could be built-in, Jerne provided brain scientists with a revolutionary concept. Fundamental to his idea is the core concept of selection and how it can be the operative mechanism at the cellular level even though it might look like instruction occurred at a higher level of organization. When trying to apply selection theory to what always appears to be a plastic brain, it is particularly important to remember what is meant by selection. To quote Jerne: "selection refers to a mechanism in which the product under consideration is already present in the system prior to the arrival of the signal, and is thus recognized and ampli-fied. Thus, at the level of the entire system: all such processes are instructive, whereas all instructive processes at a lower level imply selective mechanisms."

Jerne drew three analogies between his beloved immune system and simple learning. The immune system is forever changed by the appearance of each new antigen, just as the brain is somehow changed by the appearance of each new experience. Second, each system, brain and immune system, appears to have a memory: When the same antigen presents itself a second time to an organism, the latter produces more and better antibodies. Finally, the experience one organism has developed for its immune system is not transferable to its progeny, just as my skiing ability is not necessarily transferred to my offspring.

With these analogies, Jerne considered first a region on an antibody molecule in both mice and men called the "kappa light chain" (fig. 2.1). Each of these chains has what are called a "variable part" and a "constant part," each consisting of 107 amino acids. Over these regions of the molecule for any one species, the various amino acids that are linked

FIGURE 2.1   This crucial part of the antibody is the kappa light chain. Each chain has both a variable and a constant part. Redrawn from Roger Y. Stanier, John L. Ingraham, Mark L. Wheelis, and Page R. Painter, *The Microbial World* 5th ed. (New Jersey: Prentice-Hall, 1986), figure 30.1, p. 599.

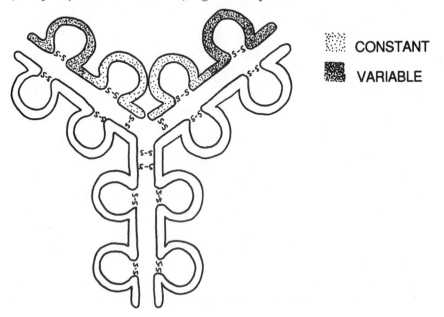

CONSTANT

VARIABLE

together to make up the region are constant for the constant region and variable for the variable region. In humans, whereas the constant part is constant for all human antibody molecules, the variable part is different for each antibody of any one person. These differences in antibodies in the human are of the same kind and magnitude as in the constant region of mice. This is much like the claim that ontogeny, or normal development, mimics phylogeny. As Jerne observes, "Phylogenic differences between species in the constant part of the light chain are mimicked by the ontogenic plasticity of the variable part."

Similarly, instincts are both managed and driven by brain circuits and fixed for any one species. However, most organisms have what looks like plasticity, and can learn, which, Jerne suggests, allows them to mimic the total of all phylogenically developed instincts of different species:

> In the immune system, the constant part of the light chain is obviously laid down in the DNA of the zygote, and it is equally clear that there is DNA in the zygote that represents the variable part of the light chain, although, ontogenically, this DNA may exhibit an immense plasticity.
>
> In the central nervous system, instincts are also obviously encoded in the zygote, most probably in the DNA. But if DNA acts only through transcription into RNA and translation into protein, and if the phenotypic expression of instincts is based on particular arrangements of neuronal synapses, the DNA through RNA and protein must govern the synaptic network in the central nervous system.
>
> (Jerne, 1967, p. 201)

In one magnificent analysis, Jerne provided both a sound biological context for how the brain circuits that enable simple instinctual behavior can be driven by genetic mechanisms and how selection can play on those circuits, and a basis for the construction of more complex and varied circuits, which would enable higher-order functions that appear to involve learning. There is no doubt that instinctive behaviors, which involve a complicated orchestration of commands to the motor system, preexist: This corresponds to the constant part of the kappa chain. What looks like plasticity in human behavior is like the variable part of the kappa chain. Yet since we know that, too, is controlled by DNA, is it not likely that our behavioral plasticity may also be driven by such DNA mechanisms?

While today no one knows for sure whether selection or instruction processes are the rule for brain processes, current knowledge about brain development does give us insight into why selection processes are the most probable mechanisms for not only how brains develop, but also for how adult brains function in an ever-challenging environment. To spell out this distinction, it is helpful to break the issue into two separate phases: The developmental phase, when the brain, and especially the cortex, is being established for its life's work; and the adult phase, when, many neurobiologists believe, the brain remains plastic at both the cellular and circuit level.

Brain Development

Since what we are ultimately after is how what we know about the function of the brain may serve concepts of selection in a cognitive sense, I will concentrate on what is known about cortical development. This is a simplification in that, while the cortex is crucial and necessary for our mental life, it is not sufficient. The cortex works in harmony with other parts of the brain that are assumed to be involved in the maintenance of body arousal levels, sleep, and other basic functions. Still, much work has been done on the cortex, as a place where there is play between internal and external events.

Early Theories: A Genetic View

The background for thinking about brain development comes from the classic work of several Nobel laureates, including my mentor, Roger W. Sperry (fig. 2.2). Sperry's work, carried out in the 1950s for the most part on fish and rats, set the stage for a rigid view of brain development. In many ways, he was the ultimate structuralist. In a series of experiments conducted over thirty years, he and his colleagues demonstrated how the nerve fibers from the eye of a fish coursed back to the brain and hooked up in a specific, orderly way to the visual lobe, a structure called the "optic tectum." The dorsal or top part of the retina projects to the bottom part of the tectum; the ventral or bottom part of the retina projects to the top part of the tectum. Fibers that were experimentally diverted and placed in the wrong part of the tectum avoided making connections, but coursed instead to their normal position. This work led to Sperry's chemoaffinity theory of development, which postulates that each nerve cell has its own specific chemical code and marker, and each is destined to establish itself in a specific fashion with target cells in the brain.

He carried out other experiments in the rat that suggested that nerve

FIGURE 2.2 Roger W. Sperry, perhaps the leading brain scientist of the twentieth century, was one of the first biologists to argue for the importance of innate and highly specific nerve nets in the brains of vertebrates. Photograph by Lois E. MacBird for an article announcing Sperry's Nobel Prize, in Caltech's *Engineering and Science,* November 1981.

fibers that connected to an "end" organ, like the muscles of the leg, are also prespecified and specific. In short, in the original Sperry view of the nervous system, brain and body developed under tight genetic control. The specificity was accomplished by the genes' setting up chemical gradients, which allowed for the point-to-point connections of the nervous system. This powerful theory is the basis for much of the contemporary work in neurobiology.

Sperry's view differed from his mentor's, the great biologist Paul Weiss, who felt that form preceded function in development. He had experimented with implanting a third leg into the frog (fig. 2.3). As this unnecessary appendage soon began to function and behave like a leg, Weiss

FIGURE 2.3 Paul Weiss promoted the resonance theory of nerve growth, which argues that form preceded function. In his view, indicated by the double arrows, nerves growing into peripheral targets, such as limbs, became instructed by the target organ itself about the kind of nerve it should be. The single arrow indicates Sperry's opposing view of specification. Source: Paul Weiss, "Homologous Response in Amphibian Limbs," *Journal of Comparative Neurology* 66 (2 [1937]).

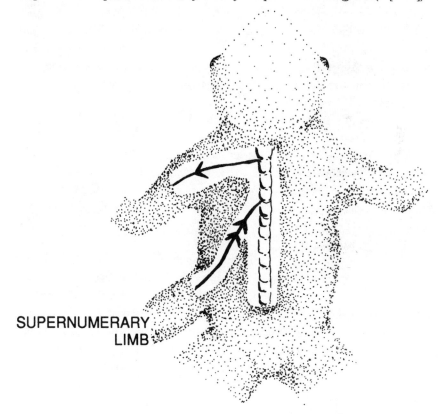

SUPERNUMERARY
LIMB

reasoned that nerves grew into the leg at random. Once there, the leg muscles told those nerves what to do, and they performed those tasks. They instructed those nerves to change what they had been doing and become leg nerves. As subsequent work has shown, neither Sperry's nor Weiss's view is strictly correct or incorrect.

In the early 1960s, ground-breaking research was also going on in the visual system of the cat. Harvard neurophysiologists David Hubel and Torsten Weisel were making fundamental discoveries about the nature and organization of the visual cortex of this beast. They discovered how brain cells are organized and interconnected to form processing units that seemed to play a crucial role in vision. They started by showing the organization of the adult cat's brain, which, in ensuing years, has been found to possess many of the same properties seen in other animals, such as monkeys and (most likely) humans.

Weisel and Hubel went on to show that the intricate organization they had uncovered in the adult cat is largely present in the newborn kitten as well. This finding, when reported, was viewed as strong evidence that the basic organization of the visual cortex is under strict genetic control and therefore not influenced by the environment. These researchers noted that the cells which they recorded in the kitten sometimes did not respond in the same robust manner as those of adult cats. In addition, they found that the orientation of the stimulus that provoked a cell to respond was less specific in the kitten. Still, the basic organization was the same for both kitten and adult cat.

Hubel and Weisel also reported a series of experiments that showed how the normal organization of the visual cortex can be modified by manipulating what the young kitten sees during development. In brief, Hubel and Weisel established that the nerve cells in a cat's visual cortex are composed of cells that respond to light presented to either eye. These "binocular cells" make up about 80 percent of the cat's visual cortex. Other cells respond only when one of the eyes is stimulated; these are called "monocular cells."

This normal pattern was radically changed if the normal visual input to the young kitten was changed. For example, if one eye was sutured closed and the other left open, the majority of the cells in the cortex became monocular. Furthermore, if the eyes were alternatively closed and opened, the same switch of cell type occurred. Here, timing of input was discovered to be important for normal development. Taken together, these studies suggested that while the main features of visual development are under genetic control, some adjustments can occur in the brain as a function of an altered visual environment. There is, thus, a phase in development where the brain is plastic and can be molded and shaped in a particular way.

Toward a Theory of Plasticity

Starting in the 1970s, the notion of plasticity began to appear in other areas of research. Investigators studying the goldfish also began to manipulate its visual system. Here, surgical removal of half the tectum revealed that the ingrowing fibers from the whole retina cram themselves into the remaining neural space and all make connections. This finding was contrary to the Sperry notion, which would have hypothesized that only the fibers that normally innervate that part of the tectum would hook up, leaving the others to simply die off. Again, the new evidence argued for some kind of plasticity.

Other evidence for the necessity of a particular visual experience for normal development came from the human clinic. After birth, the brain relies heavily on signals from the sense organs. Prior to the use of antibiotics, babies were frequently born with congenital opacity because of infection in the eye, specifically the cornea or the lens, and, as a consequence, were unable to receive sharp images. Either these babies received only blurred images, or their retinas could image only diffused brightness changes. While their retinas continued to function in the presence of unfocused diffuse light, they were not sending signals to the brain's higher visual centers in a normal way.

Years later, a surgical procedure, the lens transplant, corrected the optical defect in the eye and restored normal optical conditions. Initially, the lens transplant operation was performed on children who had been born blind, and had remained so over their first decade or more. Although everyone expected that with the new lenses these children would open their eyes and see normally, none of the patients was able to use the neural signals coming from their retinas in a normal way. They experienced the new signals, now crisply focused on their retinas, as noisy or painful. No patient was able to make use of his or her new visual information to orient in space. These patients could not learn to see, to process patterns, to recognize patterns, or to form representations of visual patterns. All of them became depressed, and some eventually committed suicide. In short, none of them succeeded in using their eyes properly.

Both animal and clinical data together indicate that normal brain development requires a key signal from the environment if normal connections are to occur. This does not mean that brain development is not driven by genetic factors. What it does mean is that the developing brain has evolved in the context of a particular environment. The genetically driven sequence of events occurs with the expectation that certain signals are forthcoming from the environment. Abnormalities in brain development that result from a radically changed environment do not speak against selection mechanisms, but merely underline the fact that a strange environment can cause serious errors in the development of a dynamic neural system. Thus, for the human, as I have suggested, the absence of normal signals from the retina allows bizarre neural connections that can result in real deficits, making it difficult, if not impossible, to see.

Yet why is this so? Why has a system developed self-organizing principles that allow transient disturbances of sensory signals to cause irreversible and severe destruction of functions? To leave the visual system vulnerable to environmental influence seems risky, and one might assume that selection measures would rid the organism of this uncertainty. But, as we shall see below, there is a considerable gain for the price that can be paid if things do indeed go wrong. Wolf Singer of the Max Planck Institute in Frankfurt, Germany, has developed a view of this issue. He convincingly argues that some brain systems, like the visual system, must remain adaptive during early development in order to adjust to changing signals received from the eyes.

Brain Circuitry and
Neuronal Matchmaking

When the embryo starts to develop, cells migrate and wander over its cellular mass. Everyone agrees that Sperry's view is more correct than not, and that it is likely that most of these cells possess chemical markers on their surface that somehow tell the cell where to stop and hook up to another cell. It has been proposed that chemical gradients guide these processes. However, as the embryonic cells continue to differentiate, some of these cells become neurons, the building units of the adult brain. When a cell declares itself a neuron, this neuron becomes electrically active, a unique property and one that is important for continuing development.

Once electrical activity is available, electrical signals are translated into biochemical signals, which can, in turn, influence gene expression (fig.

2.4). That is, electrical signals can, by having access to the information stored in the genome, influence the operation of genes. In short, electrical activity that arises out of the embryonic cells that become neurons can actually turn around and influence brain development by influencing when and where other chemical processes appear in the developing process. Singer feels this fact has three important consequences for understanding how the brain develops.

One consequence is that electrical signals can be conducted rapidly over large distances. No longer are cells merely influenced by the chemical process of the next cell or by a gradient that may pass over it. With the introduction of the capacity to conduct electrical activity, no cell is an island. Information from part of the developing embryo can reach across a great distance and influence neural development; topological vicinity and physical contact are no longer constraints for the signaling system that contributes to the structuring of the developing brain. One among many examples of such influences is that events occurring in one hemisphere can influence developmental events occurring at the same time at very remote parts of the other hemisphere.

The second consequence of having electrical signals influence the

FIGURE 2.4 The developing neuron becomes electrically active. Modern neurobiology has discovered that once a nerve becomes electrically active, it can influence the genes, which in turn influence how the nerve develops.

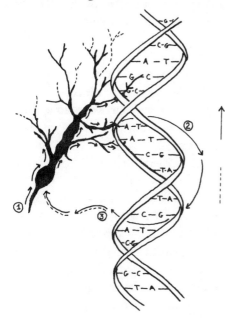

↑ PATH OF ELECTRIC CURRENT
THROUGH NEURON
TO GENE

PATH OF NEURONAL DEVELOPMENT
AFTER RECEIVING SIGNAL
FROM GENE

developmental process is that as soon as sense organs become functional, physical stimuli from the environment can interact with the developing organism and influence its development. Thus, when the eye, which is part of the brain, becomes active and can transmit information to higher centers in the brain, this information can influence, for a time, the developing dynamics of the basic neural circuits of the brain. Indeed, all sensory stimuli that start to bombard the higher brain centers can modulate electrical activity in the brain, which, in turn, can influence brain development.

The third and perhaps most profound consequence of these electrical signals is that they are the very same signals used by nerve nets to carry out the computational operations of the brain. Put differently, it is neuronal signaling in all its intricacy that solves all the decisions the brain makes, both consciously and unconsciously. As a consequence, it is reasonable to assume that developing nerve nets can use their unique ability to perform logical operations to structure their own development. In short, the brain can use its own intelligence to promote its own development. The young emerging nerve network can make the decisions about what has to be done in the next developmental step, and can do so relying upon a very sophisticated evaluation of ongoing states or environmental conditions. As the brain becomes more complex, these processes can become more sophisticated.

One crucial example Singer presents to support this view of how environmental stimuli interact to shape the development of the brain is the problem of stereopsis, or depth perception. Most animals with eyes in front of their heads have the ability to use information about the geometric discrepancies in the images presented to the two eyes simultaneously and to estimate how far an object might be in front of them (fig. 2.5). Stereopsis allows one to measure the distance of an object in space without moving the head—an important capacity if one wants to apprehend an object without moving around and altering it, especially if it is another animal, which may be a potential meal.

The second important function of stereopsis is that it allows one to measure the distance between various contours and thus helps one to distinguish a figure from its background. Thus, if you look at an object that has the same texture as its background and is 20 inches in front of it, you will not be able to distinguish object from background if you use only one eye. As soon as you open the second eye, you can use the disparate information that image produces: It immediately pops out as being a figure that is coherent and different from its background.

For stereopsis to be possible, a sophisticated circuit must be estab-

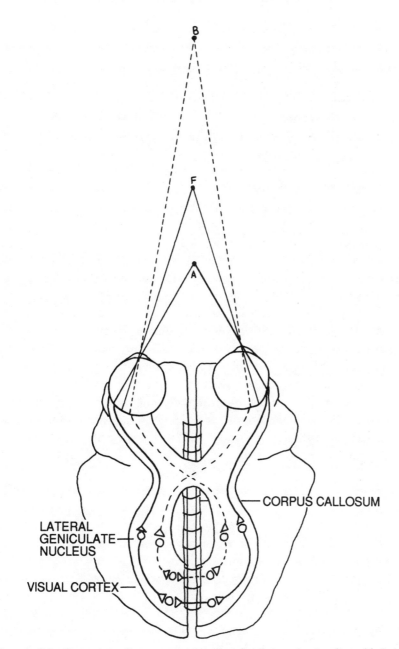

FIGURE 2.5 Stereopsis allows us to perceive depth in our visual world. It is a capacity of a special part of the brain, which proves to have an intricate and dynamic developmental mechanism. Redrawn from Eric R. Kandel and James H. Schwartz, *Principles of Neural Science,* 2d ed. (New York: Elsevier, 1985), p. 874.

lished in the brain. It is now known that it is impossible to establish such circuitry from genetic instruction alone. Physiologists have discovered that crucial and identifiable epigenetic factors play a role in the normal development of the circuits that allow the brain to perceive depth. The example described below shows how the nervous system does interact with the environment.

Each retina sends neurons to both half brains by means of ganglion cells that pass the information to a way station in each half brain called the "lateral geniculate body." The lateral geniculate body then sends information up to the visual cortex at the back of the brain, most of these neurons going to a particular area called "area 17," or the striate cortex.

In order for stereopsis to occur, not only must nerve cells in the brain gather together information from each eye, but the information from each eye must be looking at the same point in space. For points in space straight ahead of you, area 17 and its related area, 18, must get input from the other half brain. This task is accomplished by the biggest fiber system in the brain, the corpus callosum, which sends information over from the visual area of the left brain to the visual area of the right, and vice versa. A particular cortical cell receives selected input from each retina and, more specifically, from a particular and corresponding locus within each retina which encodes image points that have the same disparity. These connections have to be made for all the points in the midline of vision plus all of those points to either side of the midline, a task involving a million fibers in the left eye and a million fibers in the right eye.

As originally proposed by Sperry, this wiring could perhaps be done by chemical markers, because we know that selecting fibers through chemical gradients can be quite precise. Thus, there is no a priori reason that this connectivity mechanism could not be active in primates such as ourselves. However, frogs are one thing and higher vertebrates quite another. Other factors for more advanced organisms such as cats and primates make this mechanism unlikely.

Human beings, as well as many other animals, enjoy binocular vision, and Singer believes that its development rules out a simple chemical gradient plan for neural development. Even if there were a sophisticated chemical marking system available to sort out and label the millions of fibers involved and get them to their respective targets, how would the system know how to adjust to the several parameters of growth and development of the body?

This point is crucial because, during development, the size of the eyeballs are continually changing. The precise configuration of the retina, the intraocular distance of the eyes, and the precise rotational position of the eyeballs in the orbit are all changing in the early months of life (fig. 2.6). These parameters cannot possibly be anticipated with any great precision because they themselves are vulnerable to epigenetic accident. Also, extraordinary precision is required for this connectivity to work for stereopsis.

These problems may all be solved by the final hookup of cortical cells, a process Singer calls "functional criteria." By this process, retinal loci

FIGURE 2.6 The developing primate brain undergoes a period of growth that necessitates that the neurons responsible for integrating information from the two eyes constantly change during development. Since the head grows, the interocular distance also changes; hence, the neurons in the brain must be able to form new synapses during this critical phase of development.

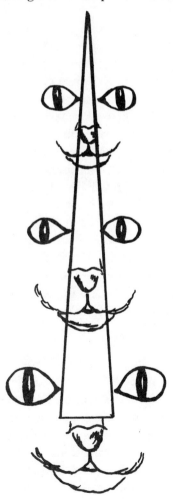

(one point from each eye that is looking at a point on an object) set up an electrophysiological pattern that must be more like those patterns set up from other retinal loci that are looking at different aspects of the object. Singer asserts that there is a probability that the activity generated by corresponding retinal loci will be more similar than activity generated by noncorresponding retinal loci—since, by definition, different loci are looking at different aspects of the visual field. Singer suggests that an unusually exuberant set of neural connections is needed to solve the selection or specification problem: That is, during development more connections are needed than will be necessary for the adult brain to operate normally. Then, through some kind of resonance-matching mechanism, a cell permanently accepts the two input cells whose resonance of pattern of firing is similar. We do not yet know whether this last appealing idea is correct.

It is now known, however, that a certain sloppiness exists in the developing brain. The chemical markers that guide the general features of the visual system leave a lot of neurons unspecified. In addition, the fibers arriving in the visual cortex from the two eyes have a huge overlap where they project in the developing animal. Singer suggests a mechanism capable of evaluating the degree of coherence of electrical activity conveyed by the neurons that carry information from the respective retinal loci of the right and the left eyes. Once it has recognized the coherence in the electrical patterns, the brain must know how to selectively stabilize those that have the highest probability of conveying correlative activity and to destabilize all of the other connections serving other retinal loci.

This wonderful bit of neural matchmaking goes on for the first three months of a cat's life, the first year of a monkey's, and the first two years of a human's. What we know about how the neural fibers connect up in the visual cortex comes from the work of Hubel and Weisel, who showed that most cells in the visual cortex are binocularly driven. Hubel and Wiesel found few cells in cat or monkey that responded to stimulation from only one eye.

They also showed that if one eye is patched during development, the distribution of cells radically changes, with most cells now only responding to stimulation of one eye. In addition, alternating the stimulation pattern to each eye also disrupts normal visual development. In fact, if input to the two eyes is switched every few hundred milliseconds, abnormal connections develop in the visual cortex, and binocularity breaks down. The developing neurons start to compete, and the cells switch to monocular arrangements. In short, Hubel and Weisel showed that normal brain development requires an exact and correlated input from each eye.

The basic anatomy and physiology suggests that each binocular cell receives multiple inputs representing all different points on the retina. Somehow, in the normal development of the ensuing selection process, this overgrowth of connections is whittled down to a subpopulation of those cells that allow for the proper functioning of a particular binocular cell. When the developmental process concludes, each binocular cell receives inputs only from cells that convey information from the corresponding points from each retina. These experiments give crucial clues about the dynamic nature of brain development and provoked Singer to offer his idea about how specific connections are established in the brain.

One of the unanswered questions in Singer's intriguing theory is how the brain knows when and when not to look for correlations in neural information. For example, if the system were looking for correlations in visual space during rapid eye movement (REM) sleep, many things would look uncorrelated, and potentially good connections would fall off. During REM sleep, noisy information would be bombarding these same visual cells and might result in false correlations. Or, if the animal is just knocking around and looking swiftly from one point to another, it would be arduous for a system to sort out which cells would hook up for proper binocular use. To avoid either of these eventualities, the system must know when to search for correlation.

First, it may be, as some behavioral studies indicate, that other brain areas outside of the primary visual system help to modulate and influence the normal development of the visual cortex. In one series of experiments, cats had one eye closed and the other eye was rotated in its orbit to see whether the distribution of cells in the visual cortex change as the cells switch from binocular to monocular. No change was found in cats whose eye had been rotated, indicating that the cells are not plastic under these conditions, even though a good deal of data implies they should be. We must conclude, therefore, that some other part of the brain modulated the normally disruptive manipulation of visual information.

Other studies show how brain centers working at a distance from the visual system influence development. In one of these studies, proprioceptive information from the eye muscles was eliminated: Thus, while an animal could move its eyes, it did not have good information on where they were in the orbit. Here again, there was no change in the visual cortex. In short, plenty of evidence suggests that there are extraretinal systems somewhere in the brain that are involved in setting the visual system for these adaptive changes.

Recent research has revealed that other brain systems might be involved. One brain system works with the neurotransmitter noradrenalin, and arises deep in the brain from the brain stem. It innervates a wide range of cells in the visual cortex. Another pathway uses the neurotransmitter acetylcholine, and arises in the midbrain. If these pathways are severed, there are no adaptive changes to visual cortex cells following eye patching. Or, if chemicals that block the actions of these two neurotransmitters are injected into the animal, these adaptive changes are blocked as well.

Singer has gone on to show a number of important phenomena that argue for the crucial role of neural activity in establishing normal connections. If a cat is anesthetized so as to prevent all of the extraretinal inputs from influencing visual cortex, no adaptive changes occur. However, if the neurotransmitters are dumped into the relevant part of the brain during anesthesia, adaptive changes do occur.

Many other events and actions must be coordinated in order for this to work. It is important for the afferent cells from each eye, the so-called "presynaptic cells," and the binocular cell (also called the "postsynaptic cell") to be molecularly ready to accept the coordinate input from the presynaptic cells. Critical molecular systems have to be ready to fire off so that when the two afferent inputs do fire coherently, the postsynaptic cell says, "O.K., let's make a synapse and be permanent." It is now known that this synergism between molecular and electrophysiological mechanisms does exist. In brief, receptors on the postsynaptic cell will open up and allow the cell to respond when there has been not only a neurotransmitter release, but also a release of a limiting magnesium ion by the postsynaptic cells itself. This ion is released only when the cell is firing at a particular frequency.

This simplified explanation may help us to understand how neural cells fulfill their destiny: An overabundance of synaptic opportunities is followed by a paring down of fibers. This paring-down process is guided in part by functional criteria that, in turn, reflect sensory signals from the environment. If those sensory signals are bizarre, the system develops abnormally and presumably puts the animal at a competitive disadvantage. When all of the foregoing is taken together, the so-called plasticity in the system is more apparent than real. In reality, the environment is selecting cells out from a larger set for a particular function.

Thus, it is clear that what is called plasticity in the brain occurs at an early point in the development of an organism. The observed plasticity, however, does not suggest that maleability in the face of environmental input defines the neural processes for a general-purpose learning machine. In fact, once again, the environment is simply signaling the developing visual system, and when signaling is normal, a certain set of connections are made. Indeed, the developing visual system is intricately designed, changing only to inputs that are relevant to setting the degrees of freedom the genes cannot set by themselves. The system is designed to be responsive only to relevant variation; it is not a general-purpose correlation soaker-upper.

THE SOMATOSENSORY SYSTEM

There are certain known general principles of cortical development emerging from current research that apply to all mammalian cortical sensory areas. They have been recently well articulated by Herbert Killackey of the University of California, whose ideas about the development of the somatosensory cortex appear, at first glance, to argue against the idea of selection.

In brief, the fine structure of the cortex develops as the result of peripheral stimuli (information) contributing to the specification, and ultimate determination, of what a particular cortical cell becomes. This rather new view is somewhat at odds with prior strong claims—suggested by Sperry and Pasko Rakic among others—that each cell of the cerebral cortex was specified through strict genetic mechanisms. Yet, the new view is no less heavily dependent on genetic mechanisms, although it does suggest an interaction between two genetically driven systems: the cortical cells themselves and the cells that synapse upon them that are transferring information from the periphery. Consider the somatosensory system of the rat.

Killackey studied the microanatomy of the part of the rat's cortex normally involved with somatosensory information, which has a relatively stable and set structure. He then compared how that structure looked if an animal had undergone amputation of a forelimb at birth. The adult cortex of such an animal looks very different, with a larger representation for the hindlimb. The interpretation is that as they make their presence known through lower brain structures, the incoming hindlimb fibers grab a larger proportion of the cells available in the cortex after amputation of the forelimb has occurred (fig. 2.7).

The reason that the cortical cells waiting for the forelimb innervation are not taken over by the equally close cells that represent the lower lip is thought to be that the lower-lip systems develop before the hindlimb system, and the afferent information from this system is already in place before the new challenge from the periphery. Yet there could be another

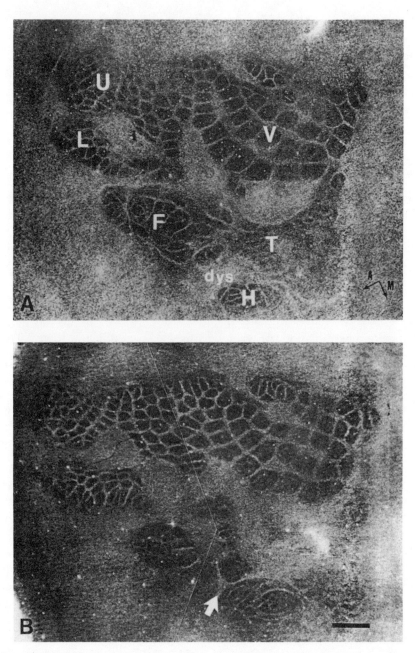

FIGURE 2.7 The rat's cortical zone for representing its body surface can change radically if damage occurs to that body surface. Here, with the forelimb removed, the part of the brain that normally represents that area has shrunk and has been taken over by the hindlimb. Used by permission of Herbert Killackey.

reason. Changes that are seen are limited to within a sensory map. Even within a modality there can be shifting around of representations within a group of related functions, but not between functionally different areas, as Pasko Rakic has demonstrated for the visual system.

Rakic showed that if one eye of a monkey is removed around the sixtieth day of development, major changes occur in the thalamic way station and in the cortex. Instead of the way station being layered and receiving inputs from each eye, the remaining eye takes over the entire way station and, in turn, sends massive inputs to the primary visual cortex. Normally, the primary visual cortex is organized into columns, with each column representing the input from one layer of the way station and hence one of the eyes, while another column represents the input from another layer of the way station and the other eye. Now, all cells represent only one eye in the primary cortex. However, even though there has been a massive change in the organization of the primary visual cortex, area 17, the next visual area, area 18, does not change at all. None of its cells move over to invade 17. The changes are within a functional group.

When, however, Rakic took out both eyes, there appeared to be plasticity once again. There were changes involving area 18 as well as other areas. Upon close inspection, such animals have abnormal connections established from other systems such as the corpus callosum. Here, fibers—or more accurately, some fibers—from the other side of the brain that normally do not innervate this primary visual area, do grow to fill the void created by the removal. While this kind of reasoning argues for some kind of plasticity, I think it is a red herring.

One always has to be wary of research that demonstrates what biological systems *can* do as opposed to what they normally do. Skin, if injured, can heal, but that is not the normal function of skin cells. Neurons committed to one destiny can invade foreign areas, but that is not what they normally do. This is really a pseudoplasticity and probably of little biological consequence. The technique of disturbing the normal course of development yields phenomena that help one to understand normal development, but it does so in an odd way.

Perhaps the most dramatic claim for the plasticity of cortical cells has been made by David Frost, now at Harvard. In the mid-1980s he literally diverted the normal retinal input from visual areas into the somatosensory cortex. Sure enough, the normal somatosensory cortex started to behave like the visual cortex. On the surface, such findings demonstrate that cortical development involves an interaction of central and peripheral systems, but there are still many issues to resolve with such experiments. Perhaps, for example, the cells that seemingly become more visual have really not changed, but, rather, the input to them has changed, and their response to that abnormal input is simply the way such a cell responds to that particular input.

Nonetheless, we are left with a very dynamic picture of brain development. At one level, instructional processes appear to be at work. Central neurons wait for information from the periphery in order for normal development to go forward. If the messages change, a different brain organization results with possible functional consequences. How does such a biological system fit the terms of selection theory?

Studies of the development of both the visual system and the somato-sensory system suggest that there is an initial biological substrate affected by environmental information. The number of possible outcomes for the final nerve nets that develop in a particular brain are determined by a selection process that involves an exchange of information between the available neurons and the environment. The higher centers of the brain, such as those that deal with visual and touch information, are not totally prespecified by genetic mechanisms, but they are established through give-and-take with the environment. Once this developmental phase is over and the circuits are established, we have to consider whether those circuits can be easily changed in the developed or adult brain.

Selection Process and the Adult Brain

In understanding brain circuitry and how it functions, scientists can only approximate how circuits work. After all, each neuron of the brain can make up to 50,000 synapses, and the human brain has over 11 billion neurons. When the possible interactions are computed it is evident that the number of possible sites of information exchange rivals the number of molecules in the universe—it is a figure that defies comprehension. Certainly in such a complicated neural space, synapses might easily change their connection now and then. If they are in fact able to change, it is likely that environmental influences could change the basic neural architecture of the brain.

On the other hand, we know from commonplace clinical examples that if damage occurs to the left brain of a small child, the brain is able to adapt, and the child can go on to use language and to speak. The same lesion in an adult greatly diminishes the capacity to adapt. This kind of observation argues against easy brain plasticity. The fully developed brain seems unable to adapt to change, the basic circuitry having already been set and established.

However, this static view of the adult nervous system has been aggressively challenged by research over the past fifteen years, most notably from the California laboratory of Michael Merzenich. He makes the intriguing claim that the cortical maps that exist for various sensory modalities can be altered in the adult by experience. Maps are, in reality, groupings of neurons in the surface of the cortex that respond to stimulation on the body surface in a way that reflects the organization of the body. Thus, for touch, the body map on the surface of the cortex processes information about touch. There is a hand area, a foot area, a face area, and so on. As I mentioned, these maps have been viewed as static systems in the adult brain.

In an early set of experiments, Merzenich carefully measured the cortical areas associated with a digit of a monkey's hand. He then amputated the finger associated with the cortical area he had measured and looked for changes in cortical representation. He discovered the cortical area shrank in size and the representation of other finger areas expanded into the space. In other experiments, where he measured the cortical area of the middle finger and then trained the monkey to use that finger in a repetitive task, he discovered that the cortical area expanded in size as if the greater functional demand resulted in greater need for cortical space. Merzenich and others have interpreted this finding as revealing that the cortical maps are in a constant dynamic relation and maintained by competitive processes generated by environmental challenges.

This basic observation has served as one of the keystones to Gerald Edelman's theory of what he calls "neural Darwinism." The issue, of course, is whether the seemingly new cortical maps reflect a change in the underlying neural structure through growth and reorganization, or whether the changes reflect a selection process where a preexisting inactive circuit is activated by a change in the environmental stimulus. At this writing, it is not known which of these possibilities is true. Needless to say, selection theorists believe that the latter possibility reflects the truth. Also, very recent work carried out by Ted Jones, a neuroanatomist at University of California, Irvine, shows there are no anatomical changes in somatosensory cortex associated with deafferentation of sensory information from the hand. This would suggest the plasticity seen by Merzenich is due to selection from other prewired circuits. Like the Rakic experiment, Merzenich's work in the adult shows plasticity only within a functional map, providing a further example of plasticity as really a selection process.

The Modern Neurobiology of Learning and Memory

Most modern neurobiologists want to believe that learning and memory are somehow carried out by changes at the synaptic level, and, like Merzenich, that learning or experience changes the very organization of the brain—a dominant idea for which the evidence is meager. A vast field of talented and busy scientists are studying the various physiological and metabolic events that can occur at the synapse, but it is not clear whether the phenomena being studied have anything to do with psychological memory. Instead of reviewing the data for the idea of synaptic change, an activity carried on regularly by the leaders in the field of memory research, I will examine a competing idea: that memory is distributed throughout neural networks in the brain. No one is quite sure what this means, but some of the data collected by adherents of this idea can be explained by selection theory.

Patricia Goldman-Rakic and her colleagues at Yale Medical School have recently come up with a riveting set of observations on the monkey. Only a few scientists have achieved the extraordinarily difficult task of correlating physiological processes with behavior, and she is one of them. In brief, she trained a monkey to look at a point in space, and then flashed lights around the point of fixation. The monkey was trained to maintain its fixation until after the light went off and the central fixation light subsequently dimmed. Then the monkey was to look at where the light had been turned on (fig. 2.8). Goldman-Rakic then recorded the accuracy of the monkey's eye movements during this task, and, after the monkey learned it, which could take months, she recorded from individual neurons in an area of the monkey's frontal lobe known to be involved with eye movements.

Her findings are clear. After the monkey had been trained to perform the task, she found individual neurons in the frontal lobe to respond in an orderly way. Many of them were firing away at their background rate until the stimulus came on. When recording from a particular neuron, and when a light came on at a particular location, the neuron would immediately fire much faster. But most important, after the light went off, and the monkey

OCULOMOTOR DELAYED RESPONSE

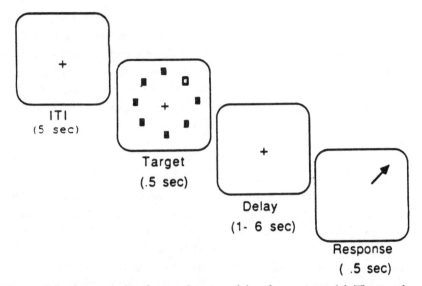

FIGURE 2.8   An example of an oculomotor delayed-response trial. The monkey is required to fixate a central spot on a TV monitor (panel 1); a target is briefly (0.5 sec) presented in one of eight locations (panel 2); a delay of 3–6 seconds commences (panel 3); the fixation spot disappears and this is the "instruction" to the monkey to move its eyes to where the target *had been*. Source: P. S. Goldman-Rakic, "Prefrontal Cortical Dysfunction in Schizophrenia: The Relevance of Working Memory," in B. Carroll, ed., *Psychopathology and the Brain* (New York: Raven Press, 1991), figure 1.

was still fixating on the central point, the neuron continued to fire. And, just before the animal moved its eyes to the light, the neuron stopped firing at a high rate. In short, Goldman-Rakic found that what appears to be a "memory neuron" is truly beautiful work. The question to be addressed now is, what does it mean?

From the point of view of selection theory, that neuron was a hard-wired unit that had been responding that way ever since birth. When the animal was "learning" the task, a task that was hardly normal in the life of a monkey, it was simply sorting out a host of irrelevant strategies and possible responses down to the one the experimenter was rewarding and, subsequently, recorded from. That neuron exists to tell the monkey where to move its eyes. The neuron has always done that, and it always will. It was built in the brain for that purpose. Clearly, from this perspective, it is not a "memory neuron": It has not somehow been trained to respond in this task-relevant way. This is the neuron's built-in function, which has evolved over millions of years.

I leave the study of brain development with a strong sense of the large involvement of genetic processes involved as well as good evidence that many of these processes play out by selection. On the one hand, it seems clear an organism would want a lot of its development driven by tightly controlled genetic processes. However, it clearly needs to adapt to many changes thrust upon it not only from the dynamics of its own body growth but also from peculiarities of the environment. The developmental process seems to be a good time for these environmental forces to play out.

We have seen how the powerful biological processes play out: Selection processes are the rule in evolution, in immunology, and perhaps also in brain development. Organisms at every level examined in the biosphere have a huge repertoire of possible responses to environmental challenge. Cells within the body are predetermined and ready to go if challenged by an intruder. An animal will or will not survive, based on its capacity to respond to an environmental challenge. And within the brain itself, selection processes appear to be actively determining how it becomes wired for adult functioning. It is time now to consider how selection might work at the psychological level.

CHAPTER 3

The Developing Mind:
The Interaction of Genes
and Environment

ONE of the attractions of the selectionist view is the insight it provides into the way the environment affects human development. Here again, scientists continue to argue about the role of nature versus nurture, of genes over environment. In the strict structuralist view of the world, organisms are largely programmed to develop in set and specified ways. From pollywogs to Polynesians, the genetic imperative marches on, independent of environmental influence. The problem always raised with this parsimonious view of development is the huge degree of variation that exists within each species. This variation becomes even larger when one considers the seemingly radical changes in body morphology, physiology, and biochemistry a *single* species can undergo when reared in different environments. Yet some of this variation is more apparent than real in that ontogenetic variation may be little more than a built-in capacity to adapt.

Stephen Gould reminds us that our extensive capacity to adapt to an environment, such as altitude, is different from the mechanism of adaptation as the term is used in evolutionary theory. The well-known physiological factors that change as a function of living at higher altitudes are all part of a cardiopulmonary system whose response range has been established by natural selection. The physiology system of a flatlander who moves to the mountains changes in a systematic way due to a plasticity that has always been part of the latitude possessed from birth by the cardiovascular system. Such physiological adaptation is not an example of a previous nonexistent capacity now appearing as the result of instruction.

Still, this kind of "adaptation" data has led researchers to reject the structuralist view that genes launch an organism inexorably on a single unswerving path of development. More specific examples of true ontogenetic adaptation occur in a variety of other systems, such as the developing visual system (see chapter 2) and in the development of capacities as divergent as bird song, social behaviors, and even human language. Additionally, coupled with the variation in all of these processes is the phenomenon of "critical periods," which has been noted for years: That is, certain environmental events must happen at certain times in the development of an organism in order for normal development to occur. Only recently has it become acceptable to admit that there must be some interaction between these two factors. Setting a new framework for that interaction is one of the benefits that a selectionist view allows. Thus, in mental development, as in the immune system, the environment can be understood as signaling the brain's many built-in circuits, including the ones responsible for psychological processes, under particular circumstances. Before examining in detail how the selection idea works in human development, consider a simpler example from neuroethology (fig. 3.1).

Peter Marler, now at the University of California at Davis, initiated a series of brilliant studies in the early 1960s. He was fascinated with how bird song develops. Does the sparrow learn its song from its father, or

FIGURE 3.1 In a classic experiment, Peter Marler observed that when white-crowned sparrows were exposed to dialects of their song, they could learn those variations but not the variations heard in other species.

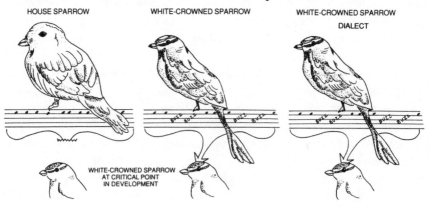

HOUSE SPARROW   WHITE-CROWNED SPARROW   WHITE-CROWNED SPARROW
DIALECT

WHITE-CROWNED SPARROW
AT CRITICAL POINT
IN DEVELOPMENT

does it simply blurt out the appropriate song? Can, say, a white-crowned sparrow learn the pattern—or more accurately, the template—of the common sparrow, the hawk, or any other species? Marler and other researchers have shown that birds must pick up their adult song pattern by listening to the pattern of an adult male at a critical point in their development. Their answer takes us back, once again, to selectionist thinking.

Marler showed that while there is some variation in what the white-crowned sparrow will learn, it is fairly limited. Thus, if the young white-crowned sparrow is exposed only to the song of adult sparrows or to no song at all, it will never learn to sing a song as an adult. If, however, it is exposed to a dialect of the white-crowned song, it will learn the local variation of the song just fine. Here again, the environment signals a built-in capacity, one that strictly limits what is possible for any given species.

Variation Among Siblings

Selection theory makes sense out of family studies that have tried to tease out the relationship between genes and environmental influences. Which of the two is responsible for the wide variation seen in behavior and cognition?

Traditionally, the studies on monozygotic twins are trotted out where it is fairly clear that genetics must account for 50 percent of the variation seen on several behavioral and cognitive measures. After all, twins reared apart differ little from those reared together on tests of cognitive skills and even on tests of personality. Dizygotic twins, while failing to correlate as highly as monozygotic twins, still correlate much more highly than unrelated control groups. Finally, tests on adopted children correlate more with their biological parents than with their adoptive parents. All in all, these findings present a strong case for the role of inheritance in these important aspects of life.

It is interesting to look at the question of inheritance the other way around, where it is clear that the remaining 50 percent is *not* due to family environment. After all, monozygotic twins reared apart have only a minutely smaller correlation on cognitive and personality measures than do monozygotic twins living together. If environmental variables were important, one ought to see much larger differences in these measures. Also, dizygotic twins reared apart have smaller differences in their correlations on these tests than do dizygotic twins living together. As for adopted children, their scores are not significantly different from those of randomly selected individuals.

Consistent with the foregoing are some new studies on siblings. Siblings have more of a common genetic history than not, but still have large individual variation on these same dimensions of human activity. Some researchers have commonly assumed that since siblings' environments are roughly the same, any variation must be due to their particular genetic component. Others argue the opposite with respect to the environmental influence.

Although untangling these complex issues looks next to impossible, two psychologists have tackled the problem with real success, and their rather startling results support a view consistent with the selection theory. David C. Rowe of Oberlin College and Robert Plomin of Pennsylvania State University first assessed the extent of true variance in the cognition, personality, and psychopathology of siblings. Not surprisingly, they observed higher correlations in all of these measures with those subjects whose genetic makeup was more similar. At the same time, siblings still have a lot of variation in cognition, personality, and psychopathology.

These studies do not indicate how much of the variation is due to genetic influence and how much to environment. The studies on monozygotic twins have forcefully argued for the role of genetics in understanding similarity. Arguing that there are now ways to examine whether genetic or environmental factors can account for the dissimilar nature of a related group of people, Rowe and Plomin have looked at the variation in sibling behavior within a family.

Indeed, every parent knows about the wide variation in response to the same event that occurs among one's own children. Yet most developmental psychologists have emphasized the common traits of a family in order to determine the best principles for child rearing. Rowe and Plomin have developed a way of looking at the differences among siblings in order to measure the role the environment plays in those differences.

Using a complex model of what they call "shared" and "non-shared" environmental factors, they determined that most of the variation seen in siblings is due to their non-shared experiences. However, in pursuing the environmental factors that may account for the variation in the non-shared environments, Rowe and Plomin have had little luck. In examining areas like sibling interaction, family structure, and differential treatment by parents, they found no clues regarding which environmental factors might be responsible for the variations seen in siblings. Indeed, it could be argued that if environmental factors are important, they are random and can never be deciphered. Thus, a pat on the head by a friendly teacher may have a huge effect on child A and none on child B because of the previous life experiences of the two children. Or, extreme heredity-environmental interactions could occur in either direction. For example, John, as a father, indulges his tall red-haired shy kids, but disciplines the short blond ones. Or, John, as a son, is challenged by parental criticism but lies around in response to praise; Bill, his brother, does the opposite. There could even be random factors such as odd experiences in the womb or slight variations in neural connections during development, and so on. Finally, all kinds of things happen in siblings' non-shared lives, such as one sibling, but not another, being bitten by a dog.

Rowe and Plomin leave us with the conclusion that parents have little influence on their children, and that genetic factors ought to be considered to explain the variation. Their message presents daunting challenges to many assumptions of child rearing, as well as to our educational system.

In the present context, however, the analysis also suggests that the framework of selection theory is consistent with their observations. As Jerne pointed out, just as each of us has a unique set of antibodies undoubtedly controlled by DNA, we may also have a unique set of neural nets, which enable us to have different capacities. These different capacities would be brought out and predicted by non-shared environments, not shared ones. Again, each child would be seen as having a vast repertoire of capacities that is largely genetically specified. Each of us has dozens, if not thousands, which we never use, but they are ready to be used in the appropriate environmental condition. This state of affairs would explain how much non-shared environments seem to affect the variation seen in siblings: Their different experiences reveal different capacities. Thus, the diversity, observed from the view of selection theory, is not surprising or peculiar, but is rather predictable.

BRAIN DEVELOPMENT AND PSYCHOLOGICAL READINESS

The foregoing analysis raises an intriguing question concerning the overall nature of psychological development. If environmental influences only signal a set of possible brain circuits that develop largely under genetic control, can environmental inputs that occur before the appropriate repertoire of circuits have developed influence development? The great child psychologist Jean Piaget realized this possibility dozens of years ago, and often argued that it is futile to expose children to certain kinds of information too soon in their development. He believed that children emerge more or less through identifiable stages during development; and that one cannot expect a child to develop, during an earlier stage, mental tools appropriate to the final states of development.

The implication here is that the developing child is gradually establishing a set of brain circuits that, once in place, can be selected by an environmental event—an assumption that studies of the developing human brain have shown to be true. Up into the teens, new cortical networks are being generated.

Piaget felt that up to the age of two children mainly acquire information about the world through their senses. The abilities to reason and to use language to assist in reasoning presumably do not emerge until later in development. For this reason Piaget felt that verbal explanations are virtually meaningless at an early age, an observation that seems to fit with every parent's experience. It makes no sense to attempt to use reason on brains whose circuits have not yet developed the capacity to carry out adult human reasoning.

There are important caveats to this line of thinking. Modern work by developmental psychologists has significantly qualified Piaget's conclusions. Renée Baillargeon, for example, has completed a series of compelling experiments demonstrating that small babies understand much more than they are commonly able to express (fig. 3.2). In light of these studies, Baillargeon challenges Piaget's argument that objects disappear from a baby's consciousness when they are placed behind a barrier.

Baillargeon had babies watch a cardboard barrier that was hinged on top of a flat surface and could be either erect or folded down. Using a well-developed method in developmental research she moved the barrier up and down until the baby tired (adapted out) of the experience and looked away to other events in the room. Then while the barrier was in the up position, Baillargeon slid an object that had been previously in the baby's view behind the barrier. As predicted, the baby looked away as if the object was outside of consciousness. When Baillargeon moved the barrier part of the way down, so that it appeared to hit, and appropriately stop at, the still-hidden object, the baby did not pay a significant amount of attention to her gesture. However, when she moved the barrier all the way down, so that the barrier appeared to move *through* the object, the baby quite dramatically became alert and looked right at the barrier. In other words, the infant did not express surprise when the barrier failed

FIGURE 3.2 Babies remember the presence of an object after it has been removed from their vision.

to pass through a hidden object, but became very surprised when one object seemed to pass right through another.

This extraordinarily clever experiment shows that, contrary to Piaget's theory, a baby does have the mental representation of objects, even when they are not in full view. The brain knows the lawful relations of such events and does make the appropriate calculation. Once again, we have a strike against learning and in favor of built-in capacities.

This general idea has received much support from animal studies. Research on the development of swimming in amphibians has clearly demonstrated the uselessness of sensory experience. Tadpole larvae have been raised in the presence of drugs that inhibit neural activity all the way up to full development, thus preventing sensory information from being conducted into the developing nervous system. As soon as the drug is removed, however, the animal is immediately able to swim. In short, early sensory experience does not prevent the tadpole from swimming.

Some researchers have argued that the phenomenon of "infantile amnesia" relates to this issue. It has been well known for years that the long-term memory of young children is poor. They seem not to build up memories in their earliest years, as if the organism needs only short-term memory for their day-to-day management. If the memory mechanism of the young child is different in nature from that which develops later in life, it is a waste of time to try to build into the young child's brain experiences that he or she will not remember.

All of the findings suggest that it is futile to teach very young children cognitive skills. The yuppie parent who wants Johnny to read at age two is not only trying to drive a biological system that isn't ready for the task; he or she may be creating emotional conditioning that will have unhappy consequences for the child in later life.

When the bumps and grinds of early development are settled, we still have to understand that remarkable machine, the developing brain. Specifically, we must comprehend, in terms of selection theory, how young children acquire such vast amounts of information. Needless to say, it is a matter hotly debated in cognitive neuroscience. The philosopher Daniel Dennett lamented a few years back that psychologists will be "passing the buck to biology." He suggests, "We will have to accept the disheartening conclusion that a larger portion than we had hoped of learning theory . . . is not the province of psychology at all, but rather of evolutionary biology at its most speculative." Dennett has complained that selection theory has gone overboard and cannot believe that learning is not a product of instruction. He has a lot to be concerned about, especially with respect to new work on how the organism learns.

Constraints on Learning

True learning would appear to reflect how individuals develop specific skills. After all, one learns languages such as French, English, Italian, or Chinese as well as activities such as tennis, golf, and so forth, all seemingly from scratch. Yet a major topic of study in psychology has been the issue of whether the newborn infant has a clean slate; or whether constraints, built in by evolution, determine what an infant is able to "learn" about his or her environment. This issue is central to any understanding of mind building.

Even the most ardent behaviorist, like John Watson or B. F. Skinner, always worked on the principles of learning. In trying to specify how reinforcers have to be presented to an organism in order to maximize learning, they and their droves of disciples would come up with relationships between rewards and behavior that they would then call a "principle." Of course, any such relationship would mean that certain principles were already built into the organism.

Constraints on learning have always captured the attention of linguists, who—as members of one of the major disciplines examining development—also have the most to say about the issue. They have known for years that there is a significant amount of nonsense in the environment to which children are subjected. Parents, for example, do not reinforce their children in a way that would encourage them to use language correctly. For example, when a child announces, "I went poo-poo on the floor," parents tend to become upset, and do not praise the child. Yet the child has uttered a perfectly correct sentence. Conversely, when the same adorable language machine says, "Daddy home?" the parents coo in response to the ungrammatical utterance. How is a naive child supposed to sort out the essential principles of semantics and syntax from such random kinds of reinforcement and social circumstances?

It used to be argued that children were born with some general-purpose inductive machine that allowed them to sort things into categories. From Jean Piaget to Jerome Bruner, the idea was that a child, upon encountering an object, immediately and reflexively generated a hypothesis about it; then subsequently tested the hypothesis against other examples and gradually determined into which category the object should be placed. This view was abandoned as it became known that children have a hard time rejecting negative evidence for a hypothesis. Yet, from eighteen months on, they are learning new words at the rate of nine items a day. What kind of brain machine would allow for that phenomenal rate of learning?

In the early 1960s, the philosopher William Quine further complicated the understanding of learning by pointing out the problem of inferring induction mechanisms. Since the list of possible hypotheses for what a particular new object might be is almost limitless, and since children learn language at lightning speed, there must be some built-in mechanism that makes a child decide which hypothesis to seize upon when confronted with a new object. Once again, the specter of the child being born with millions of years of evolutionary experience packed into its mental apparatus looms large.

In fact, the word has come from Noam Chomsky and his disciples that children don't learn language, but are born with the understanding of how the system ought to work. Thus, when a teacher says, "This is a truck," children understand that the label refers to the truck and not to its color, a particular wheel on the truck, the action of giving the object to the child, or a host of other possible meanings. A naive learning machine having no constraints on what constitutes an input would have a devil of a time sorting all that out. Yet children know immediately the label refers to the truck. How can this be?

This sort of realization has led several researchers to suggest that, when learning language, the child has a built-in hypothesis about the salience of objects. In other words, upon hearing the word "truck," the child focuses attention on the whole truck and assigns the name to the whole object. The child does this because millions of years of evolution have built this into the child's brain. One doesn't learn the rule; one does learn "truck." This capacity has been described as the child's ability to make the whole-object assumption, and reflects an early and primitive feature of the language system. In addition, after the child has assigned a label to an object, other associations fall into place—as if learning a word tended to catalyze the whole cognitive system.

Normally, working without language "tags," children appear to have a very different cognitive structure than when learning the name of something. For instance, when given pictures of items to sort or pair together, children tend to carry out the chore "thematically." Thus, children will place the picture of a cat with a dish of milk as opposed to placing it with a picture of a dog (fig. 3.3). They do this with great regularity and, indeed, this pairing makes complete sense.

Now, if children see the world in this contextual way, why don't they assign a label to these themes as opposed to distinct categories? That is, why don't they assume "cat" means "cat drinking milk"? Enter a second constraint. Several years ago, George Miller of Princeton University pointed out that following object assumption, the young brain then also says the object is a "type of" something. Ellen Markman at Stanford University has recently investigated this idea in terms of what she calls the "taxonomic assumption." When learning a word, children switch their attention away from looking at themes and direct it to the object; they then inductively develop ideas about the category into which it must fit. All of this goes on automatically, as the language system kicks in to label things.

Markman went on to confirm these ideas in a series of ingenious studies and demonstrated the central importance of what the introduction

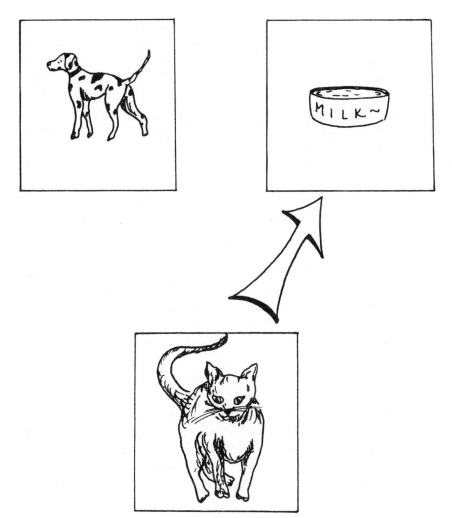

FIGURE 3.3 Young children, not yet aware of the names for items, sort objects thematically; thus, the cat and bowl belong together. As the child begins to grasp categorical relationships, the choice would switch to the dog.

of a word does to the cognitive system. In a series of tests, Markman tried to see whether children organized objects by category (for example, milk-juice) or by thematic relation (for example, milk-kitten) when the objects were presented with either a novel object label or no label at all. The children presented with objects that had no labels chose the categorical relation only 25 percent of the time. When the target object was labeled with an unfamiliar word, children matched the target with a category member 65 percent of the time. These findings were replicated when pictures of artificial objects were used instead of real objects. In both cases, children focused on categorical relationships simply because of the presence of the word and not because of any particular knowledge about its meaning. In short, children place an abstract constraint on what single nouns might mean and implicitly reject hypotheses about category membership as a result of having learned a label for the object. Several other findings support this conclusion.

Markman discovered further examples of the automatic process at work. In addition to developing categorical relationships about objects in the presence of a label, she also found that children constrain word meaning further by assuming that words are mutually exclusive—that each object has one and only one label. In other words, familiarity with objects seems to dramatically affect the exclusive nature of children's labels. For example, when either a metal cup (familiar) or a pair of metal tongs (unfamiliar) was described as "pewter," children were likely to call a set of wooden tongs "pewter" (7 out of 12), but unlikely to call a ceramic cup "pewter" (1 out of 12). Thus, if a novel label is applied to an object for which children already have a label, they reject the new term as an object label.

These studies indicate that young children may be innately constrained to consider single nouns to refer to objects of the same *type,* than to ones associated by theme or event. This innate knowledge helps explain how children acquire new words as rapidly and efficiently as they do. By constraining the meaning of a term to categorical relations, children are able to rule out a large number of other potential meanings for any given term. Although this ability provides a critical first hypothesis about word meanings, children must eventually be able to learn terms for properties of objects. The mutual exclusivity principle can be used to constrain successively the meanings of terms.

Category Formation

Once categories are formed, there is good evidence that very young children use this information to make inferences about the nature of their world. This is true despite widespread assumptions that children cannot form mature categories or look beyond the obvious physical features of objects. There is recent evidence from Susan Gelman of the University of Michigan that young children can use category membership to predict underlying similarities among objects even when perceptual similarity would lead to a different prediction. The fact that children work from categories in something like the biological sense suggests that members of a category have a hidden underlying "essence," allowing children to make predictions about physical similarities rather than the other way around. Turning the question around, Gelman considered how children limit their inductive inferences from categories. Previous work had fairly well established that important category constraints on induction include homogeneity and the naturalness of a category. When categories are highly homogeneous and clearly full of natural items, induction is fairly easy.

In one study, Gelman gave an induction task to two age groups—preschoolers and second-graders. For each of a series of problems, the children were taught one new fact (for example, This rabbit likes to eat alfalfa) and had to decide whether the fact applied to other objects different from the original object (for example, Do you think this telephone, cow, and so on, likes to eat alfalfa like this rabbit?). Gelman varied the categories by including both objects occurring in nature (natural kinds) and objects made by humans (artifacts), and by how similar (homogeneous) category members were to one another.

Gelman discovered that all children place constraints on their inferences. Even preschool children expect certain categories to have a correlated structure that goes beyond appearances. Second-graders also relied on category level to guide induction. Beyond such similarities, however, marked differences were evident between preschoolers and second-graders. According to Gelman, second-graders make more distinctions among categories. Their real-world knowledge feeds into the primitive categories young preschoolers naturally seem to have.

All of this, of course, invites consideration of the parallel that language learning and concept formation seem to have with the immunologic mechanism of selection. In this light, the ability to form inductive hypotheses about the world seems to be constrained at the very start by the limitations involved in learning word meanings. Children appear to be innately endowed with the capacity to order objects taxonomically when a single noun label is present. Here is an example of complexity built into the brain, complexity signaled by an environmental challenge that appears to be an instructive process. Further, this built-in taxonomic categorization capacity brings along with it many other capacities that enable a series of inferences, such that properties that apply to one member of a category can be assumed to apply to all members of that category. The acquisition of more information with regard to word meaning or to the scientific nature of the object fine-tunes the categories. New information is processed within the parameters of preexisting category structures. Thus, "constrained learning" seems to parallel the theoretical conception of immunological selection.

PERSONALITY DEVELOPMENT

Learning a language is complex, though somehow seemingly more manageable than determining whether selection theory applies to our own personality structures. Personality seems so pervasive throughout our conscious systems, and so influenced by our environment, as to make remote the idea of selection theory being appropriate to this dimension of our lives. Yet a good argument can be made for selection theory in this domain, even with respect to the powerful phenomena of critical and/or sensitive periods.

A good deal of the personality research on infants and children has focused on temperament. Although the theories of temperament differ in specific ways, they generally consider the content of temperament to include behavioral style, reactivity, emotionality, activity level, sociability, and impulsiveness. The gist of these studies indicates that temperamental dimensions appear early in development, are relatively stable over time, and show evidence of heritability. This stability has led some observers to conclude that temperament may be considered the foundation for later personality.

Some particularly interesting work has been done on racial/cultural difference in temperament. In 1976 D. G. Freedman, for example, was able to demonstrate clear differences between newborns of Chinese ancestry and matched groups of Caucasians on temperament items from the Cambridge Neonatal Scales. Although the two groups were indistinguishable in many areas of development, European-Americans tended to reach peaks of excitement sooner and to show greater instability to new states. Chinese-American neonates, on the other hand, were calmer, steadier, faster to habituate or accommodate to external stimulation, more highly consolable when upset, and better at self quieting. Similar evidence for this racial/cultural difference is supplied by Harvard's Jerome Kagan.

Of course, one of the problems encountered when studying infants for personality or cognitive development is the ambiguity concerning the degree to which their prenatal environment has influenced subsequent neonatal behavior. Adding to the unknown in this time frame are the influences that may occur during the so-called "sensitive periods" of development.

Sensitive periods refer to those times during development when environmental events may have a disproportionately powerful influence on the organism. Investigators of these periods have been reluctant to acknowledge an underlying unified framework for them, and have tended to concentrate their descriptions to particular, individual periods. Leaders in the field, such as Marc Bornstein of the National Institute of Child Health and Human Development, feel that researchers have consciously eschewed generalizations, with the result that the overarching issues of the structural character and causal interpretation of sensitive periods have been insufficiently addressed.

Still, most investigators believe that sensitive periods are standard to early development, and much sensitive-period research and theory are predicated on these opinions. However, sensitive periods in infancy have been studied most often. Exceptions show that sensitive periods can also arise late in development. In addition, although sensitive periods are thought normally to occur only once in the life cycle for any one modality, some sensitive-period alterations in structure and function can take place more often.

Most researchers agree that the onset of a sensitive period is rapid, but full susceptibility is not always or necessarily reached at onset. For example, as I described in chapter 2, while kittens that experience monocular deprivation at the start of the sensitive period may lose binocularity, loss at that time appears to be only transient, because binocular experience in the balance of the sensitive period can lead to significant recovery. In contrast to the speed of onset, the end of a sensitive period is typically gradual: That for human binocular vision is believed to taper off over about five years.

Our knowledge about the effect of the environment on the sensitive period and the specific pathways by which it operates is far from complete. Although neurotransmitters regulate the temporal frameworks as well as the sensitivities to external stimuli basic to virtually all sensitive periods, no one has pursued the particulars of this process. It is also not known whether experiences in the sensitive period induce, attune, or maintain an outcome. Richard Aslin pointed out in 1981 that experience may induce or attune development at the neurochemical level, whereas experience may maintain development at the physiological level. Questions about outcome conditions of the sensitive period usually include when and under what circumstances in development the environment influences both the onset of the sensitive period and the experience itself. The rules we do have point to the fact that experience plays a particular role in sensitive periods, but does not specify what that role is.

A typical example of how an environmental event can influence brain development is now plastered on the wall of every bar in America and will soon be on every beer can as well: Alcohol use during pregnancy can affect the fetus. The influence of alcohol is particularly damaging at certain times during pregnancy, and alcohol use during pregnancy is considered to be a major cause of mental retardation in the Western world.

Alcohol has different influences at different stages in development. In the first trimester, alcohol influences the development of bodily organs; in the second, it can influence brain growth; and in the third, the development of neural circuits. Whether these influences are substantial depends on many factors, including, of course, how much alcohol the mother consumes.

One of the brain structures that can be influenced with alcohol use is the corpus callosum. This structure interconnects the two cerebral hemispheres. Normally the callosum sends out more fibers than it needs to adequately connect the two hemispheres. During the usual course of development, many of these extra cells die off with no noticeable loss of function or behavior. With alcohol in the picture, however, the cell death seems to be abnormally high. And, in fetal alcohol syndrome, the callosum can be underdeveloped or not present at all.

Yet an interesting aspect of this syndrome suggests that selection mechanisms might be a factor. Specifically, we know that some neural connections will die off in the presence of alcohol, and that others will not. In this light, it is possible that some neurons are more susceptible to threatening environmental agents than others. This variation in susceptibility would suggest that some brains exposed to alcohol may be influenced by its presence while some might not—as has, in fact, been observed and reported by Ernst Abel. A pair of dizygotic twins had a different response to their mother's drinking, one of them developing fetal alcohol syndrome while the other did not. Here, both fetuses were exposed to the same factors, yet only one succumbed. It would seem that the developing neuronal circuits in the one twin were more directed and resistant than those of the other.

We are still far from knowing exactly how the environment acts on the individual, and how innate constraints affect this interaction. Studies in personality, temperament, and cognitive development consistently point to the importance of incorporating models of biology in their conceptual and methodological framework, as well as the necessity of specifying the interaction between the environment and the individual. While this evidence does not conclusively support the argument of selection as a mechanism of psychological development, it does support the notion that pursuing selection-based models is likely to prove fruitful.

CHAPTER 4

Language and Selection Theory

THERE are over four thousand languages in the world today, and thousands more have been lost over the course of time. Of course, there are even more dialects, as well as variations in formality and style within each language. When Noam Chomsky surveyed this field of variation, he noticed a certain unity and made the dramatic proposal that all languages share some common characteristics. Since then, we have learned that all languages do have many things in common. It follows that human beings must have brain mechanisms that support our basic capacity for language and determine the commonalities. Still, this claim rattles our very perceptions of the rich complexities in spoken languages. How can there be such order in this endless variation? How can it possibly be driven by a common "selected for" capacity in our species brain?*

---

*Incredulity on this point comes from two giants of evolution and linguistics theory: Gould and Chomsky. In one form or another, both have argued that language is not the product of natural selection, claiming that it tags along on the back of some other evolved entity. In this position, they are not unlike Eisenhower who, though he had on his side power, resources and truth, inexplicably stopped his march toward the Soviet Union at the end of the Second World War—a decision that had grave consequences for the future. In not taking their fundamental insight all the way home, Gould and, especially, Chomsky left the field open for some misguided speculation.

Pinker and Bloom, recognizing this situation, moved in to help fix up matters (fig. 4.1) by placing language back into the biological world, and attempting to illustrate that understanding the evolution of language is no different from trying to understand how stereopsis evolved. Without a doubt, the mysteries of language capacity can be overpowering to those who have not dissected it. Thanks to Chomsky and legions of others, language has been analyzed. Now, sound insight into its structure will make for solid thinking regarding the manner in which it was selected.

Figure 4.1 Steven Pinker, linguist and cognitive scientist, has grasped the importance of viewing mental phenomena in an evolutionary perspective. Photo copyright 1987 Marvin Lewiton. All rights reserved.

When selection theory was applied to the problem of language in the past, there were poorly substantiated claims about children who spontaneously learned to speak some foreign language: That is, just as there are antibodies ready to meet any environmental challenge, there are built into all of us brain circuits ready to produce English, Russian, French, Swahili, and so on. From this idea, it was an easy step to believe that a child living in Nebraska might accidentally turn on, say, the Finnish circuits. In fact, such claims were made, but when they were tracked down, it always turned out that everyone had forgotten about the Finnish au pair who lived with the family a few years before the remarkable event. Other investigators examined the "paranormal" claims of the spontaneous use of a foreign tongue, like Bulgarian, in an adult "channeler." In each case, the claimant was stringing together foreign-sounding nonsense syllables like, "ishta vishta roveshta." While all of this is great fun, the serious business of understanding language and selection theory is far more riveting. To understand how language might have been selected for, we must learn something about its nature. In the following section, I draw on Steven Pinker's short characterization of the main properties of language. His analysis at our Venice meeting was spellbinding, and I summarize it here.

After reviewing the general properties of language that hold across all human languages in all cultures, some specific mechanisms of grammar are discussed. It is in the detailed analysis of grammar that evidence is provided for the deep biological nature of language and for how selection played out its hand. A further glimpse of the biological roots of language comes when examining how children learn language. Nowhere does the old idea of a *tabula rasa* fall on its face with such force. Children reveal to us that huge aspects of mind are built-in.

GENERAL PROPERTIES OF LANGUAGE

Language learning is universal: All children learn to talk. Language is also universal across all social classes, whatever prejudices one may have about the regional or cultural dialects of one's own language. The language of subcultures that might be considered working class or lower class is every bit as complex as the language of linguists themselves. For all people speaking all languages, the underlying mechanism is the same. Furthermore, unlike such innovations as the alphabet or agriculture, which have a definite origin and a chain of transmission, no human population has ever lacked a language. There are no reports of a newly discovered band of hunter-gatherers who lacked language, only to suddenly learn it from their more technically advanced neighbors.

The absence of class distinctions within a language also applies between languages. Indeed, the languages of hunter-gatherer populations, or of so-called other "primitive people," are remarkably elegant. For example, Cherokee, one of the North American Indian languages, has sixty or seventy different pronouns instead of the half dozen or so we use in English. Instead of a single word "we," there are separate pronouns for me and you; me and some third party; me, you, and some third party; me, you, and several third parties; me and several third parties; and so on. Likewise, there are many forms of "you" and "they." Nevertheless, children quickly pick up this marvelously specific and complex system, and use it without error.

Language is learned early in life. Starting at about one year of age, children begin to speak their first few words. They typically start to combine words at around eighteen months. By the time most children are three, they can engage in conversations and are clearly speaking English or French, or whatever their native language is. They accomplish this without formal instruction, although every culture has folk theories about how children learn language. Deeply intrenched in our culture is the myth that parents teach their children. Needless to say, children don't get formal, explicit instruction on the kinds of things they unconsciously know by the age of three. Furthermore, a surprising fact is that children are not systematically corrected for grammatical errors. When parental expressions of approval or disapproval are tabulated in terms of whether a child's utterance is grammatical or ungrammatical, it is found that parents tend not to make corrections about grammar. Nor do they misunderstand their children more often when the children speak ungrammatically. In most situations, parents are pretty much in tune with what their children want. In order to communicate their needs, children really don't need a language as much as one might think. Most of the grammatical errors children make don't impede communication at all. After all, there is nothing ambiguous about "he comed into the room." In fact, in some cases child language is less ambiguous and more communicative than adult language. For example, whereas the adult form *hit* is ambiguous for present or past tense, the child's erroneous *hitted* is clearly a past-tense form of that verb.

There are also anatomical correlates of language in areas of the brain (fig. 4.2); typically, areas in the left hemisphere seem to be specific for both language and certain aspects of grammar. Broca's area, the most

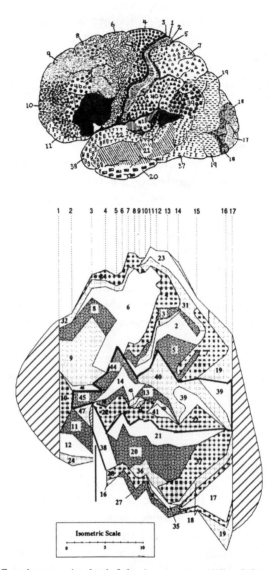

FIGURE 4.2 Certain areas in the left brain seem specialized for various aspects of language. The classical Broca's area is crucial, as is the classical Wernicke's area. My laboratory's recent development of a method to "unfold" the cortex and to observe adjacency in the various areas showed that the parts of the brain involved in language are adjacent to each other. The numbers refer to the various cortical zones as originally described by the great German anatomist Brodmann. Reprinted from Michael S. Gazzaniga, "Organization of the Human Brain" *Science* 245 (1989): 947–52. Copyright 1989 by the AAAS.

prominent, is commonly implicated as the area of the brain most critically involved in the production and perhaps in the comprehension of grammatical sentences. Lesions to this part of the brain leave patients able to understand language but usually unable to produce or comprehend the syntax of a statement. In patients whose two brain hemispheres have been surgically divided, only the left hemisphere is capable of using syntax to aid comprehension. In some of these patients, the right disconnected hemisphere can understand simple language but is unable to use syntax. Despite various controversial claims in the 1960s that other species could be taught systems very similar to language, most people would now agree that those experiments did not truly teach the equivalent of the human language to chimpanzees. This is also true for the experiments that purportedly taught language to other species, like dolphins and gorillas.

Not only does a special brain anatomy support human language, but there are also special anatomical features of the human vocal tract that are necessary if spoken language is to develop. The human vocal tract has a very sharp right-angle bend and a relatively low larynx compared with those of other primates. This configuration creates a rather large space where the tongue can move along two degrees of freedom and define two resonant cavities simultaneously. By filtering and shaping the sound coming out of the throat, these cavities produce enhanced frequencies or formants. These frequencies are then carried by the speech wave to the ears of the listener to communicate in parallel the vowels and consonants being spoken (fig. 4.3). This arrangement allows for extremely rapid communication in terms of bits per second. One estimate is that we can communicate easily 25 phonemes per second. If each phoneme were a discrete tone—that is, if the system had only one degree of freedom— then speech sounds would pretty much merge into an imperceptible buzz (as they do for some people after damage to the auditory cortex). Fortunately, however, the human vocal system is well designed for rapid communication of information.

This anatomical specialization exists, however, at a cost for non-linguistic physiological functions. Unlike in other primates—whose larynx is higher in the vocal track, allowing them the capacity to breathe and drink at the same time—every piece of food, every bit of liquid that human beings swallow must pass over the opening of the trachea. This is why adult humans are at acute risk for choking. Yet for the first three months of life, infants have a more generalized primate vocal tract and young babies are able to breathe and drink at the same time. It takes several months of development for the anatomical changes necessary for human language to occur.

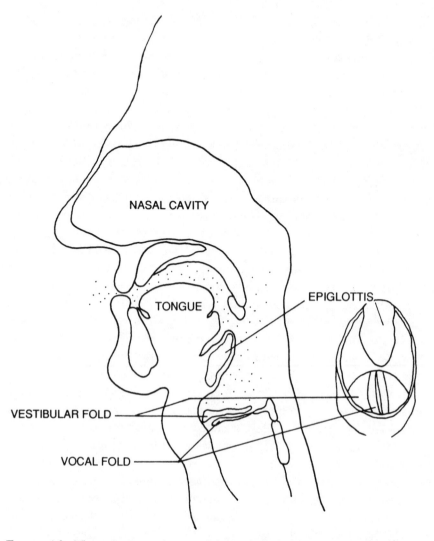

FIGURE 4.3 The unique vocal tract of the human allows for the production of the sounds of speech. Complex arrangements of the lips, tongue, and teeth, in conjunction with the controlled passage of air across the vocal chords as they expand and contract, give rise to the multitude of sounds humans are capable of making. Based on plate 73 in Wynn Kapit and Lawrence M. Elson, *The Anatomy Coloring Book*. Copyright © 1977 by Wynn Kapit and Lawrence M. Elson. Reprinted by permission of HarperCollins Publishers.

There seems to be a critical period for language acquisition, usually thought to be between the ages of eighteen months and six years. Recently Elissa Newport of the University of Rochester reported a critical case. She studied deaf children who were not exposed to sign language during their development. Their parents were affected by teachers and therapists, who, as part of the oralist tradition, considered it undesirable to teach children to sign because speaking is more accepted in the hearing world. The oralists were taught to lip-read and to speak, a very laborious enterprise that causes many deaf people to fail to learn any language. If these deaf children reach late adolescence without any language, they may actively seek sign language. When they are exposed to sign language at that time, they can learn it. Yet, when the sign language of people who have learned language as adults is compared with the language of signers who have learned sign language as children, there are noticeable differences in the extent to which their communication follows the rules of American Sign Language. Newport also found, along with Judith Johnson of the University of Virginia, that immigrants speak English perfectly if they arrive before age six, regardless of the number of years spent in this country. The case of Henry Kissinger comes to mind. He and his younger brother came to the United States at the same time, but whereas Henry speaks with a pronounced German accent, his brother's English is flawless.

Other lessons about the critical period come from the neurologic clinic. Children who suffer major brain damage to the left hemisphere early in life can adapt and go on to learn a fairly complete language, presumably with their remaining intact right hemisphere. This capacity for recovery wanes by puberty. If the left hemisphere is damaged in the adult, there is often little or no recovery. Additionally, in split-brain patients, the disconnected and usually language-impoverished right hemisphere never seems to acquire language. Although some split-brain cases have language in their right hemisphere, it is manifestly present immediately after their disconnection surgery; it does not develop later.

In the animal literature, there is now a rich body of knowledge about critical periods for development, as I already described in chapter 2 with respect to certain visual events. Other mental functions have also been observed to occur during specific times in early development in order to ensure their proper development.

While it may be tempting to think of language as a mere problem-solving skill, or as some kind of fairly simple mapping between thoughts and words, making its acquisition a puzzle that we solve easily using problem-solving strategies, there is a good body of evidence to suggest that language can be dissociated from general-purpose cognition. For example, it is easy to show that you can have cognition without language. Animals, infants before they have a language, adults who grew up without sign language, or stroke victims who are aphasic, can all display some intelligence without the benefit of language. Looking at it the other way around, patients who become demented are frequently able to speak relatively normally, but have little capacity to carry out the simplest problem-solving behavior.

Still more evidence of this type comes from Richard Cromer, who documents the case of a teenage girl in England who was severely retarded because of spina bifida. Yet to say her language was unimpaired would be an understatement. Sounding like an Oxford don, she spoke impeccable English with an aristocratic British accent, and made no grammatical errors even though her speech was absolutely devoid of content. Her language was all fantasy and confabulation.

Grammar itself may be autonomous from the general ability to communicate information. The sentence Chomsky used to illustrate this point—"Colorless green ideas sleep furiously"—is obviously different from a string of words like "Furiously sleep ideas green colorless," even though neither communicates any meaning. Conversely, word strings like "car, crash, hospital" communicate a message rather clearly, but do not follow the grammatical pattern of English. More subtly, "The child seems sleeping," is not a grammatical sentence, but communicates an idea perfectly.

These broad features of human language have been further broken down and extensively studied. Up to this point, we have seen that there is strong evidence that language has a major biological component: Specific parts of the brain are involved in its operation; there are critical periods for its development; there is a universal grammar across all languages; and so on. Even with this knowledge, however, we are still left wondering how human language became established. Did the whole entity arise out of one big genetic mutation, or did it arise gradually through selection processes as Pinker and Bloom believe?

Their argument depends on the intricate analysis of grammar by modern linguists. It is because of this complexity that some linguists, like Liz Bates at the University of California at San Diego, exclaim:

> What protoform can we possibly envision that could have given birth to constraints on the extraction of noun phrases from an embedded clause? What could it conceivably mean for an organism to possess half a symbol, or three quarters of a rule? . . . monadic symbols, absolute rules and modular systems must be acquired as a whole, on a yes-or-no basis—a process that cries out for a Creationist explanation.

As Pinker demonstrated at our Venice meeting, grammar is complex, but need not have arrived all in one lump. Just as the aphasic patient and the developing child can make their intentions known, early humans could have evolved through various stages of capacity. Indeed, the recent work of Myrna Gopnik supports this view as well. She studied a family that is unable to form plurals of words, apparently as a result of a particular genetic disorder. The grandfather, father, and sons all have this abnormality in the presence of an otherwise normal language system. Gopnik's findings would suggest that the establishment of language capacity in humans is governed by several genes, just like the development of the eye. Mutations to one gene can cause a deficit in one grammatical dimension of a language without affecting others.

MECHANISMS OF GRAMMAR

The grammar a language scientist like Pinker studies is not the grammar most of us learn in school or look up in style manuals. In traditional grammar we are certainly taught not to say things like "The child seems sleeping" or "dog big the." The grammar written in a grammar book stifles our natural tendencies and attempts to impose on us someone else's idea of correct language (fig. 4.4). In other words, our school drills in grammar are truly directed at inculcating a standard prestige dialect to be used in formal writing. Some rules taught in grammar make no sense from a linguistic point of view, and are often based on ham-fisted analogies with what scholars once felt to be more perfect languages, such as Latin. The proscription against split infinitives, for example, which few people actually follow in speech, was based on the reasoning of grammarians a couple of centuries ago: That is, since the infinitive cannot be split in Latin (because it is a single word), it should not be split in English.

Modern grammarians try to do something quite different. They try to describe the mental knowledge structure you have when you know English as opposed to when you know French. They see grammar as a way of mapping meaning onto sound for production in speech and vice versa for comprehension. In approaching this issue, we must understand what a "meaning" is, and what aspects of meaning language is able to convey.

Language has complex devices dedicated to communicating. The most salient type of information communicated by grammar is predicate-argument relations, the grammatical clues that tell us who did what to whom. For example, in the sentence "John ate the red apple that I bought," the verb "ate" is a predicate and it has two arguments—the agent, John; and the patient, the apple—which must be kept distinct. With the verb "eat,"

81

FIGURE 4.4 Noam Chomsky, the man who changed forever how we think about grammar. Photo by Christopher S. Johnson, from Justin Leiber, *Noam Chomsky: A Philosophic Overview* (Boston: Twayne Publishers, 1975).

the relation between two arguments is pretty obvious. In the case of "kiss" or "touch," you have to distinguish between John kissing Mary and Mary kissing John. Typical predicates in a human language talk about relations of actions, causation, location, and motion (such as "John went to the store," "John is in a chair"), and possession ("John has a book"). Furthermore, for any one of the arguments, there is usually some additional information expressed by modification. Thus, we comment on the fact that it wasn't just an apple, but a red apple. As a result, the sentence quickly has a topology. If John ate the red apple, it is important that "red" be specified as pertaining to the apple and not to John or to any other word in the sentence.

There is also a way of restricting the set of items about which you are talking, which Pinker loosely calls "quantification." In his example, "the red apple that I bought," he is basically saying "the apple, such that, I bought the apple"—a fairly complicated, though common, logical construct for which we use relative clauses or questions signaled by words beginning with "wh": Thus, "Which apple did you buy?" means "Which *x, x* being an apple, did you buy *x?*" Language has special devices to communicate simply such logical complexity.

Certain types of information do not pertain to any particular participant in a sentence but rather have the whole sentence as their scope. Tense deals with the relative time of occurrence of the referent of a proposition, compared with a speech act, and in turn relative to a reference point. In English, tense is a system for specifying temporal relations among three elements: the event itself, the speech act, and the reference point. Most markers for tense also say something about "aspect," or how an event being talked about is distributed in time (that is, whether it is instantaneous or protracted). For example, compare these sentences: "When I walked into the room, John had broken the plate" and "When I walked into the room, John was breaking the plate." The aspect marking on the verb can either zoom in on the internal structure of the event or shrink it to a point, and thus allows you to locate it relative to another event, usually the moment of saying the sentence.

Modality is the dimension of grammar that refers to whether something actually happened, might happen, or must happen, according to the speaker. Modal auxiliaries like "can," "might," "should," "shall," and "must" express this aspect of grammar. Then, the phenomenon of "illocutionary force" refers to what you are doing with a particular proposition—whether asserting it to be true, questioning whether it is true, or commanding someone to do something.

These are some of the most salient aspects of meaning that can be communicated through grammar. One way of looking at language is as a solution to the problem of how to take one of these levels of communication (which has a multidimensional topology), and encode it into a linear channel so as to get it into someone else's head (fig. 4.5). Grammar is a device, a way of giving a standardized code, to that kind of information.

Consider predicate argument relations—that is, information about

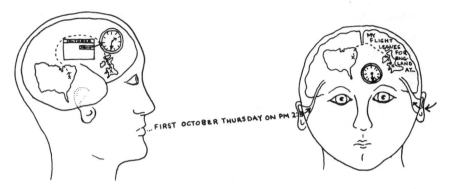

FIGURE 4.5 The speaker has a multidimensional image of what he wants to say and wants to communicate this image to his friend. His image has to be serialized into speech sounds arranged in such a way that the content of his image can be transmitted to his friend. The rules of transmission that allow for his friend to hear and decode the string of sound is grammar. Every language has its own grammar at one level of analysis. But as linguists have shown us, there is a deep structure to the grammar that is common across all languages.

who did what to whom. Argument structure is the most central component of meaning. For example, consider a predicate like "eat," with its two arguments—an agent and a patient. One way of representing it in the mental dictionary is to list it as a two-place predicate, requiring an agent and a patient, and to say that, in active-voice sentences in English, agent is encoded as a subject and patient as an object. The rest of the grammar tells you how to recognize a subject or an object when you see it in terms of the word-by-word composition of the sentence.

With phrase structure in English, the subject is defined as the first noun phrase in a sentence. This rule says that a sentence consists of a noun phrase subject, followed by a verb phrase. A verb phrase consists of a verb followed by a noun phrase. A noun phrase has an optional determiner, such as "the," "a," or "some"; an optional adjective; and a noun. While an oversimplification, this explanation allows one to define hierarchical structure—"John ate the apple"—with determinate positions for subject and object. These can serve as pointers between syntactic positions and semantic arguments. Thus, you know that in "John ate the apple," John is doing the eating because "John" is in the subject position and the lexical entry for "eat" tells you that the subject is the agent argument.

Case systems allow you to do the same thing in a different way. Although case markers don't play a big role in English, in Japanese you can attach a marker to the end of the word to indicate its grammatical role in the sentence. In this method, grammatical information becomes glued onto each word and states "I'm a subject" or "I'm an object," depending on the suffix. Thus, the word order in such a language can be varied without obscuring the message.

In expressing grammatical relations by agreement, Pinker observes that one basically builds an entire sentence around a verb by using a series of affixes to encode the verb's relations to the other parts of the sentence. A verb may have several affixes in a language that uses this system. Each position or slot in the complex word corresponds to one of the grammatical roles; and a particular affix allows one to communicate different types of generic information about that role. For example, the subject affix "ca" might denote that the subject is adult, male, human, singular, and definite. On the other hand, a different affix in that slot would have communicated female, human, nonadult, and so on. An object marker might say, for instance, that an object is inanimate, small, three-dimensional, and solid. Many North American Indian languages use this scheme.

In using these three basic mechanisms, languages can vary considerably. Obviously, the order of sentence elements can be important. Languages can be classified according to the order of subject, verb, and object that predominates: thus, English is an SVO language; Japanese is SOV; Irish, VSO. Also, there are differences in the extent to which configuration is used. English, which relies heavily on word order and uses few case markers, is at the extreme end of a continuum. Although word order plays some role in all languages, there are languages in which the role of word order is minimal, and words can be ordered almost at random. In the Australian aboriginal language (Walbiri), word order can be scrambled. You can take the adjective out of the noun phrase and put it farther along in a sentence. This can happen only in a language with rich case markings that help the listener figure out what word that adjective is intended to modify. Languages like this are at the other end of the continuum from English.

Another striking difference in language is the phenomenon called "ergativity"—a case whose distinctive function is to indicate the agent or instrument of an action involving something else. It is not immediately obvious that in English the subjects of a transitive verb and an intransitive verb are treated the same way—as in "He touched the apple" and "He ran." Not all languages work that way, however. In some languages, especially Asian and North American Indian languages, the object of the transitive verb is treated in the same way as the subject of an intransitive verb—as in "He touched the apple" and "Ran him." Most European languages don't follow this pattern.

English is one of many languages built largely around subjects, a so-called "subject-prominent language." In some other languages, however, the subject is a less important construct than the truly important topic, or what you are talking about. In some Asian languages like Chinese, every sentence does not have to have a subject, but it must have a topic.

There are also classifier systems that have a fairly rich set of descriptors for kinds of objects. In many languages, you can't talk about a stick or a chair, but instead talk about a "piece of stick" or a "piece of chair," the same way we would talk about a piece of fruit. A piece of fruit doesn't necessarily mean a piece taken out of a fruit; it can merely be a linguistic convention for a single fruit. The use of "piece" tells you that it's a discrete object. You can't say "a piece of water." Languages that take this system to an extreme have ways of classifying objects in terms of solid or liquid or number of dimensions, such as human, nonhuman, animate, inanimate. Such dimensions can also be attached to verbs in order to communicate information. This is not done without some ambiguity; but in context, the meaning is usually clear.

Chomsky has argued for years that, underlying apparent variation, all languages use the same machinery. He claimed that if linguists wrote the best possible grammars, those grammars would clarify the elements common to all languages. It would then be apparent that the mechanisms of one language can be found in some form in all other languages.

The problem of case illustrates Chomsky's point. English has a full-fledged case-marker system, though it is not prominent. In the pronoun system, "I," "me," and "my" are different in that they represent the nominative, the accusative, and the genitive cases, respectively. This pattern runs all the way through the pronoun system. In the genitive case, whole noun phrases sometimes have a case marker as opposed to only pronouns in the other cases. English also has an agreement system: "He walks," but "I," "you," or "they walk." The "-s" ending is required for third person singular only. While this sort of appendix is not used as much in English as in other languages, it clearly differentiates meaning: Thus, in the sentence fragments "the boy and the girl who walk" and "the boy and the girl who walks," we know that in the first, both the boy and the girl are walking, while in the second, only the girl is. So, in English, agreement rules do have a function, however reduced it may be compared with some other languages. The mechanism underlying the use of case or agreement—that is, the kind of rule we have to write to capture it—is present in all languages. Yet some languages "select" one as a crucial mechanism, whereas in another that feature may be only an appendage.

English is a configurational language in that it has very limited free word order. Nonconfigurational languages have a variety of markers that allow free scrambling of word order. Although this may seem exotic to us, nonconfigurationality does occur in the English prepositional system. Prepositional phrases can be strung together, scrambled in order of appearance, and still convey unambiguous meaning because the preposition, although a distinct word, acts like a case marker in denoting the role of the phrase. Thus, "The package was sent from Boston to Chicago by Mary," or "By Mary from Boston to Chicago," or "To Chicago from Boston by Mary" can all be used without altering the meaning. We have no trouble processing these sentences.

Traces of ergativity can even be found in examples like "I broke the glass" and "The glass broke," where the object of the transitive ("broke the glass") is the same as the subject of the intransitive ("the glass broke"). In fact, the rationale behind ergativity is that there are roughly two kinds of intransitive: fixed intransitives which express action, such as "I ran," and changes of state, like "The glass broke." In fact, semantically, what is happening to the glass as an object and what is happening to the glass as a subject are rather similar. Ergative language generalizes from that similarity so that it applies it to all intransitive verbs whether they imply the same kind of change of state or not.

The "as for" construction is a nice example of how to denote topic position in English. But, in English, it isn't used in every sentence. It is easy to imagine using the sentence "As for fish, I like salmon," but how about a language in which every sentence would have to have start with "As for x"? That would be a topic-prominent language.

There is even classification at work in English. Although it may seem strange to think of marking nouns and verbs in terms of coarse properties like solid/liquid, one dimensional/two dimensional, or rigid/flexible, English verbs do indeed contain such properties extensively. The same distinctions made in Native American languages are also found in English verbs. "I folded the string," sounds a little strange to most people because the verb "folded" requires a two-dimensional surface. "I plucked the bark from the tree" sounds odd because "plucked" requires something attached to a one-dimensional object. "I splashed the sand" sounds strange because "splashed" implies that the object is liquid. These kinds of distinction—dimension and rigidity—are coded in classifier languages.

So despite widespread surface variation, a look at the underlying mechanisms of case, agreement, and classifiers demonstrates that while some languages emphasize certain devices more than others, the repertoire of devices from which they choose seems to be fairly restricted compared with the surface diversity.

Of course, this hints at some rule for selection with a fixed set of devices from which to choose. There are also devices that not used by any languages, such as left-right inversion. In no language is a question posed by taking the left-right order of the words and repeating them back in right-left order. In no language is the middle word removed and put at the end of a sentence. In no language is there classification by features like color, man-made versus natural, hot versus cold, and so on. The set of devices that a language can use is restricted.

Finally, we are still left with the question of how grammar is acquired by children. Linguists like Pinker maintain that this process is a strong instance of the induction problem already mentioned in chapter 3. A child hears some finite sample of sentences from parents, peers, or siblings and must generalize from them to the infinitely large language. Yet, the child doesn't know whether that infinitely large language is English or Walbiri or Russian. The reason the task is difficult is that there are an infinite number of incorrect hypotheses that are completely compatible with all that a child actually hears: for example, a language like English but one in which you can say "the child seems sleeping," just like "the child seems asleep." When linguists realized this, they realized that children cannot be general-purpose logic machines when it comes to framing and testing hypotheses about language. Their hypotheses must be constrained in some way.

However, critics of the idea that language—which is to say, grammar—grew out of selection pressures point out the apparent diversity of devices for communicating information. The observation that, while some languages use word order for communicating who did what to whom, others use case, agreement, or some other device, has led critics like Gould to the view that some general-purpose device for complex computations evolved and was adapted, like his spandrels, for language use. As only Gould could put it:

> I don't doubt for a moment that the brain's enlargement in human evolution has an adaptive basis mediated by selection. But I would be more than mildly surprised if many of the specific things it now can do are the product of direct selection 'for' that particular behavior. Once you build a complex machine, it can perform so many unanticipated tasks. Build a computer 'for' processing monthly checks at the plant, and it can also perform factor analyses on human skeletal measures, play Rogerian analyst, and whip anyone's ass (or at least tie them perpetually) in tic-tac-toe.

Another critic, the ingenious David Premack socks it to evolutionists on the problem of language complexity:

> I challenge the reader to reconstruct the scenario that would confer selective fitness on recursiveness. Language evolved, it is conjectured, at a time when humans or protohumans were hunting mastodons. . . . would it be a great advantage for one of our ancestors squatting alongside the embers to be able to remark: "Beware of the short beast whose front hoof Bob cracked when, having forgotten his own spear back at camp, he got in a glancing blow with the dull spear he borrowed from Jack"?
>
> Human language is an embarrassment for evolutionary theory because it is vastly more powerful than one can account for in terms of selective fitness. A semantic language with simple mapping rules, of the kind one might suppose that the chimpanzee would have appears to confer all the advantages one normally associates with discussions of mastodon hunting or the like. For discussions of that kind, syntactic classes, structure-dependent rules, recursion and the rest, are overly powerful devices, absurdly so.

But as Pinker and Bloom point out, the computer that does the mailing won't type a letter unless you change the program: "Language learning is not programming; parents provide their children with sentences of English, not rules of English. We suggest that natural selection is the programmer."

But cavemen hardly needed the sort of complex grammar we academics use every day. Pinker and Bloom point out there may be all kinds of reasons that a reproductive advantage was conferred on early humans for acquiring such devices. There is all kinds of evidence that early humans had home bases and wandered about to seek out food, only to return home for its consumption. As Premack has eloquently shown, humans are the only species that engages in pedagogy. For one caveman to instruct a colleague about where food was in the space and time around the camp, grammar would be essential. In short, early humans would have found grammar a crucial advantage in pursuing all of these activities.

Consider also all of the social interactions of early humans. There is good evidence that both *Homo habilis* and modern but primitive hunter-gatherer societies engaged in complex social interactions (fig. 4.6). The !Kung gab all night long at their fires, communicating complex emotional and social feelings. In his studies of these primitive humans, Melvin Konner finds their social and cognitive interaction as complex as an Oxford don's. Furthermore, in communicating their beliefs, desires, concerns, and feelings, they need recursions, as Pinker and Bloom point out.

FIGURE 4.6 The !Kung have served as a laboratory for what may suggest examples of early human social interactions. Photograph from *The Dobe !Kung* by Richard B. Lee, copyright © 1984 by Holt, Rinehart and Winston, Inc. reprinted by permission of the publisher.

Recursive rules are not overly powerful grammatical devices at all; they allow for the human condition.

Finally, our ancestors' social interactions gave rise to other activities, such as deceit. Establishing quid-pro-quo relationships and suggestion mechanisms for enforcing them have led some investigators to suggest that there was a sort of cognitive arms race during our evolution. Deceit in one caveman encourages reciprocal deceit in the caveman's neighbor, which promotes more distrust, and so on. While coming in out of the rain may not take much brain power, figuring out how to outsmart the sneak who lives in the next cave requires some intelligent functioning. Indeed, one of the big events in human cognitive development occurs when we learn how to mask our true perceptions.

CHILDREN AND ABSTRACT CONSTRAINTS

When all of this is taken together, Pinker proposes a model for instruction versus selection using as analogy the tailor's chore: In the instructional model, the tailor makes a suit to fit, whereas, in the selection model, the client goes to a large warehouse and picks out one that fits—first filling out a large form that asks about color, size, arm length, and so on, with each question on the form possibly having a subquestion. So, at each choice point there are a small number of alternatives, yet, over all, a fairly large number of decisions. Here again we have an example of something that looks like instruction at the macro level, but is, in reality, selection at the micro level.

The same sort of mechanism may operate when one learns a word. To begin with, a lexical entry is not simply a pairing of meaning with sound. One must know the syntactic category: that is, whether the entry is noun, verb, or adjective. Then, there are subcategories. If it is a verb, it must be either transitive or intransitive. Next, the morphological aspects of the word have to be considered. What is the root? What prefixes and suffixes are present? There is also the phonology of the word. Where a word was once thought of as a string of sound segments, now it is thought of as a multidimensional structure, each dimension linked together by a series of timing slots like pages held together by a spiral wire. As a word is spoken, a variety of units must fit into these slots. The elements that must be considered include vowels, consonants, syllables, tones (if the language is a tone language such as Chinese), and so on. These elements, in turn, can be broken down further: For example, a consonant can be voiced or unvoiced. The place (labial, dental, velar, et cetera) and manner (stop, fricative, liquid) of articulation must be specified. Likewise, vowels can arise from the front or back of the mouth, or be high or low, round or nonround. The phonological form might be relatively simple to fill out because of the finite number of well-specified choices. Recent work in phonology seems to be arguing that at the microlevel, a selection process is at work.

To take it a step further, at the level of lexical semantics one can talk about an inventory of meaning elements. For example, verb meanings are built around things like basic kinds of relations, such as acting, having, going, being. Other distinctions are between things, like an event versus something that is timeless. Still, other basic elements might be things that cause, allow, prevent, and so on.

Take the verb "to butter." No one would argue that the concept "to butter" is part of a small inventory from which all language learners select.

91

The word meaning here is specific to our culture. But if the verb "to butter" means "to cause butter to be on," then the concept "to cause x to be on" may well have been built out of a small set of recurring elements that exist across all languages.

One kind of evidence for this type of selection would be the great specificity in the absence of environmental input. Put differently, is there any evidence that children structure lexical entries around this kind of analysis? An experiment addressing this issue comes from Peter Gordon at the University of Pittsburgh. He examined morphological structure to see how words are built from simpler parts, in accordance with an idea put forth by linguist Paul Kiparsky.

There are lots of ways to form words from other words—for example, adding an affix to "random" forms "randomness"; or two words can be combined to form a compound like "Nixon-lovers." Kiparsky maintains that all of these word-generating mechanisms belong to three well-defined subsystems called "levels." The first, Level 1 processes, can act on the root word. These are processes that can change the phonology of the stem; that is, they can alter the root. The application of these processes is unpredictable, and nothing about the phonological or the morphological form of the root determines whether they can be applied—"randomness" versus "randomicity" is an example. They are also unpredictable in the way they affect meaning: thus, while "reddish" is "like red," "bookish" is not "like a book."

One suffix attached by Level 1 processes in English is "-ian." When attached to "Darwin," it yields "Darwinian." Products of Level 1 can be fed into Level 2 processes where there is predictability in meaning. In English, examples include the compounds, such as "Nixon-lover," or the "isms," like "communism." Finally, at Level 3 are inflectional processes which add the regular endings "-ing," "-ed," and "-s." Kiparsky argues that these processes are ordered, and that while a word can progress upwardly from Level 1 to 3, it cannot go the other way. This predicts the patterns of formation of many words. Thus, "Darwinian" can be formed by Level 1 processes. Similarly, "Darwin" can go to Level 2 and become "Darwinism" or go from Level 1 to Level 2 processes and become "Darwinianism." But it can't become "Darwinismian." In short, the system seems to be organized hierarchically and unidirectionally.

Peter Gordon looked at interactions among four rules. Irregular plurals like "children" and "mice" reflect Level 1 processes. These irregular plurals demonstrate stem changes, and the occurrence of words that change like this is not predictable. Compounding is a Level 2 process, allowing the formation of words like "dishwasher" and "Nixon-lover." Regular plurals yielding "boys" from "boy" are the result of Level 3 processes.

To examine how these rules behave in children's language acquisition between the ages of three and five, Gordon constructed a game in which the children were shown a puppet that was doing something like eating rice. They were then asked what they would like to call the puppet. They would quickly respond that the puppet was a rice-eater. Gordon then cleverly varied the activities of the puppet to see what children would call it with roots from different levels. Kiparsky's theory was borne out by their choices: In forming the Level 2 compound, the children, having access to both roots and Level 1 irregular plurals, said "He is a mouse-eater" or "He is a mice-eater"; but, blocked from use of the Level 3 regular plurals, they never said "He is a rats-eater."

An instructional device for such an abstract constraint is unlikely because it would mean that children would have to listen for the existence of plurals inside compounds and note that they were always irregular, never regular. To check out this possibility, Gordon looked at word-frequency data, examining how often such compounds occur per million words of text. Even though there are no grammatical restrictions against irregular plurals and compounds (you can say "teethmarks"), they are extremely infrequent in the language. This means that while children might hear "dishwasher" or "toothbrush," they rarely or never hear compounds such as "teethmarks" that would give them some evidence that irregular plurals could be used in compounds. Nonetheless, they knew that it was possible, and they freely used irregular plurals in compounds, while avoiding all regular plurals—even creating their own inventions, like "mouses": They would never say "mouses-eater" even if they said "mouses" in isolation.

In this example of what Chomsky originally called the argument from the poverty of the stimulus, children are respecting a distinction even though there is little or no evidence for it in the stimuli that they hear. This type of argument led Chomsky to develop his strong views on the innateness of language. The lines of research that come out of developmental psycholinguistics all point to the importance of innate properties. The more one learns about how children acquire language, the more one comes to realize the power of the genetic aspects of language processes.

The loftiest mental capacity of all, the ability to communicate, to map sound to meaning can be explained and viewed as involving selection. Concept- and language-specific circuits resulting from thousands and thousands of years of evolution are available to humans. As environment interacts with a brain with these capacities, the system responds in an orderly way, storing the information that will guide the circuits into action specific to that environment.

Chapter 5

Specialized Brain Circuits: Selecting for Intelligence

T HE issue of human intelligence and its variation from one person to another must be seen in relation to the larger issue of our place in the framework of biology. Indeed, a full understanding of human existence requires a theory of brain function that explains what it is about all members of the human species that makes them so unerringly inventive compared to other living animals. For there is a difference between species intelligence and intelligence within a species. The latter variation in humans has come to be measured and described as an "intelligence quotient" (IQ), while the former concerns the basic principle of brain organization that must exist between species (as between apes and humans). In some sense, the between-species phenomena ought to be easier to identify.

The overall model of selection theory suggests that a repertoire of neural circuits have come to exist in the human brain and that these circuits are "selected out" for use when dealing with particular environmental challenges. This simple idea implies that there ought to be impinging on us information and circumstances that are difficult to learn very well. Evolutionary processes have probably not conferred on the human being all possible capacities to learn all kinds of strange things. There must be phenomena in the world that totally elude us because we do not have a capacity for considering their meaning. As we shall see, this is exactly the case.

Clearly, one of the hallmark features of our species is the ability to adapt over a wide range of circumstances. Consider the screen door problem. No chimpanzee in the world could instantly solve it, yet any normal member of the human species, their arms loaded with boxes, can, upon confronting a closed screen door, extend the pinky of one hand, insert it into the handle and open the door. Any member of our species can instantly rearrange items placed in a car trunk or other limited space in order to increase its capacity. In short, it is the everyday things we do that make us different from other animals. The value of these everyday and highly useful capacities for survival are not lost on evolutionary processes. Selection goes to work and finds all members of our species accumulating these life-enhancing skills.

Now, there must be something in the human brain that allows for these abilities. It is well known that the human brain is inordinately large after being corrected for body weight: Allometric considerations find the human brain falling off the correlation line assessing body size and brain size (fig. 5.1). This gigantic biological organ, weighing between 1,100 and

FIGURE 5.1   In allometric relationships for three primate species, humans show a much higher encephalization. Reprinted by permission of the publishers from *Ontogeny and Phylogeny* by Stephen J. Gould, Cambridge, Mass.: Harvard University Press, Copyright © 1977 by the President and Fellows of Harvard College.

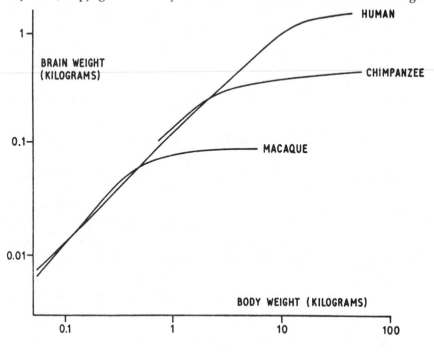

1,300 grams, sits magnificently on top of a small body, guiding its sensations, actions, desire for reproduction, and so on. These elements of human activity are indeed similar to those seen in both near and distant evolutionary relatives, and share many common mechanisms. Even basic dimensions of mental life, such as memory, attention, and visual perception, also seem to share mechanisms and brain structures in common with many animals.

And yet, why is the primary visual cortex of a human several times the size of the monkey's? Abundant psychophysical measurements have demonstrated time and again that the monkey's visual system has virtually the same characteristics as the human system, indicating that a relatively small visual system (cellwise) can do everything a large system can do. Why do humans have all those extra cells? It has always been commonly assumed that they are present to serve our greater intelligence. After all, with respect to higher-order processes, such as problem solving and language, humans simply put matters into second gear and lose the competition. The human being possesses unique cognitive capacities. Scientists have tried, over dozens of years, to liken higher processes between animals and the monkey or even chimpanzee, and fail miserably. When it comes to simple measures of intelligence, humans stand by themselves in the animal kingdom. How are their extra cells helping to serve this function?

These general issues converge on some simple questions. Does the human brain enable superior mental capacities by virtue of its larger size or by the fact that it has specialized circuits for superior mental capacities? Are the specialized circuits in such numbers that they represent all those extra cells? At the functional level, does the human possess wholly different processes than our closest animal relatives? Do these differences reflect differences in cell number or in circuitry or both? In attempting to address these issues, I draw upon a variety of recent experimental evidence that suggests the human brain has many unique features.

Selection theory would want to see evidence for our uniqueness in human brain circuitry, not in cell number, unless the latter simply reflected more specialized circuits; modern data certainly suggest, as I will show, that specialized circuits located in particular areas of the human brain enable our superior intelligence. Selection theory also argues for exclusions—mental operations that we cannot perform. I will also review the considerable number of things we humans are poor at doing.

HUMAN BRAIN STRUCTURES

As I have discussed, the human brain has two halves: the left, specialized for language and speech, and the right, with some specializations as well. Each half cortex is the same size and has roughly the same number of nerve cells, and both cortices are connected by the corpus callosum. The total cortical mass is assumed to contribute somehow to our unique human intelligence. What do you think would happen to your intelligence if the two half brains were separated? Would you lose half your intelligence since the part of the brain talking to the outside world would lose half of its support staff? Such surgical interventions (called "split-brain surgery") are carried out on patients who suffer from epilepsy, and I have been studying them for years.

A cardinal feature of split-brain research is that following disconnection of the human cerebral hemispheres, the verbal IQ of the patient remains intact (fig. 5.2). Indeed, the problem-solving capacity remains unchanged. While there can be deficits in recall capacity and on some performance measures, the overall capacity to carry out problem solving seems unaffected. In other words, isolating essentially half the cortex from the dominant left hemisphere causes no major change in intellectual

FIGURE 5.2 The brain on the left is intact, its left dominant hemisphere free to call upon the cellular activity of the the neurons of the right brain. The brain on the right has been divided for medical reasons, leaving the right hemisphere no longer in touch with the left; yet the left has not lost its ability to carry out general problem solving.

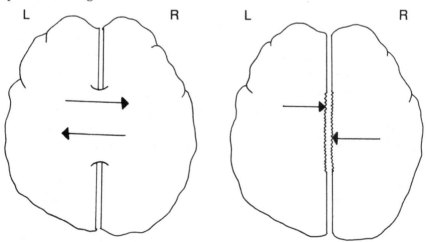

L   R   L   R

function—strong evidence that simple cortical cell number cannot be related to human intelligence.

The notion of special circuitry is supported by a vast number of observations of patients with focal brain disease, as well as by a host of studies from split-brain patients. For example, most disconnected right-hemisphere patients are seriously impoverished in a variety of tasks. While the right hemisphere remains superior for some activities, such as the recognition of upright faces, some attentional skills, and perhaps some emotional processes, it is poor at problem solving and numerous other mental activities. In fact, it would appear to be inferior to the mental capacities of a chimp. Right hemispheres lack such high-end capacity as the ability to solve problems, which the chimp possesses. We do not know whether the circuits that enable the chimp brain to carry out such operations are in this species' left hemispheres; but we do know that a brain system that has more cells than the overall organizational plan of a chimp brain—that is, the disconnected right hemisphere—cannot perform these tasks.

If we accept the view that the human brain has special circuits, what levels of organization might they affect? The overall plan of the mammalian brain has been commonly observed to be quite similar across species—primate and human brains. One of the reasons for comparative studies is the belief that homologous brain structures may carry out common functions in primates and humans. Yet, whether this assumption is justified is another matter. For example, in the brain's commissure system (the nerve fiber system that connects the two sides of the brain together) there is a particular region, the anterior commissure, that is apparently common to both the monkey and the human. Careful examination of the system, however, reveals that it is responsible for different things in each species (fig. 5.3).

Magnetic resonance imaging (MRI) technique has also allowed researchers to sharpen their hypotheses about the commissural areas so critical for interhemispheric communication in the human brain. Prior to the early 1980s it was necessary to depend on surgical notes to determine the actual extent of callosal surgery—that is, how much of the callosum was severed, and exactly where the incisions were made. With the advent of MRI, however, it became possible to confirm the extent of surgery much more clearly. Comparing the MR scans to surgical notes, we now know that estimates by surgeons of the extent of their surgery can be in error. For example, after presumed full callosal surgery on four patients, we observed that two could transfer visual information, while two other

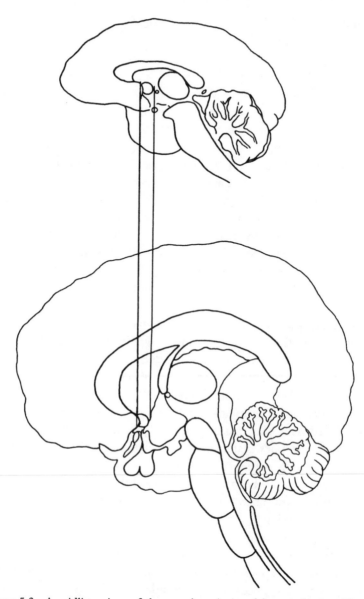

FIGURE 5.3   A midline view of the monkey *(top)* and human brains. The large nerve fiber system that connects the two half brains is the corpus callosum. The much smaller system is the anterior commissure. The callosum is five times larger in the human brain while the anterior commissure is the same size. Sources: For monkey midline view, Michael S. Gazzaniga, *The Bisected Brain* (New York: Appleton-Century-Crofts, 1970), p. 149, reprinted by permission of the author; human midline view adapted from Arthur C. Guyton, *Basic Neuroscience: Anatomy and Physiology* (Philadelphia: W. B. Saunders, 1987), p. 11.

patients did not. In an effort to account for this, we proposed that the remaining anterior commissure might vary in the kinds of information it can and cannot transfer.

When we looked at the animal literature, there were indications that the anterior commissure could transfer visual information. Based on this, it made perfect sense that the same structure could perform the same function in humans. Although only the callosum was found to subserve interocular transfer in cats, the anterior commissure was found to be involved in visual transfer in chimpanzees. This was strong evidence that the same might be true for humans, leaving the homology assumption intact.

The MR scan has allowed us to test the homology assumption: Indeed, the MR image tells all. In applying it to one of those supposedly split patients, I found that a part of the callosum had been inadvertently spared, thus explaining the transfer of information. In any such patient, in fact, any evidence of transfer of visual information indicates sparing of the part of the callosum that transfers visual information. This suggests that the anterior commissure, which is clearly able to transfer visual information in monkey and chimp, does not do so in the human.

Only a few facts are known about this interspecies difference in function. In humans, the anterior commissure represents approximately 1 percent of the neural fibers that contribute to intercortical communication, compared with 5 percent in the rhesus monkey (fig. 5.3). Thus, the human anterior commissure is proportionately much smaller in comparison with the corpus callosum and the extensive cortical territories it serves. In the monkey, between half and two thirds of the temporal lobe transfers information to the opposite hemisphere exclusively via the anterior commissure. In the human, while the distribution is not fully known, it is thought to be much less. Thus, the anterior commissure appears to have undergone substantial remodeling during the phylogenetic evolution of mammalian species.

The clear difference in function between monkey and human brains, combined with possible new anatomical correlates supporting these differences, suggests that similar brain structures across species do not necessarily perform the same function. It suggests that the arguments against such cross-species comparisons are as crucial today as they were when originally argued years ago.

The fact that monkeys and humans share many more structures with different functions, along with other, more recent structure/function research, supports the conclusion that the human brain has its own special circuits.

SPECIFIC CIRCUITS FOR INTELLIGENCE

New evidence comes from studying the cortical maps of monozygotic twins. Such twins look alike, talk alike, behave similarly, think similarly, and so on. Are their brains alike? Normally there is great variation in the gross morphology of the brain. While all brains have a similar overall plan, they vary tremendously in details. Some brains, for example, have bigger frontal lobes than others. The pattern in which the cortex appears in each of us is called the "gyral/sucal" pattern. It varies considerably, and that variation is assumed to reflect differences in underlying brain organization. The fact that the overall cognitive skills of monozygotic twins are more alike than those of mere siblings would suggest their brains are physically similar to one another.

Until recently, no one had any information on this crucial point. Now, through the technology of magnetic resonance imaging, it is possible to produce clear images of the living human brain that allow us to make the necessary comparisons. Our laboratory has been working on ways of quantifying these images in a way that would permit us to examine various regions in each half brain and to assess their similarity in surface area. We make some fifty images, or "slices," of the brain, then reconstruct them to make maps of the human cerebrum. With these maps, we can easily measure the cortical areas of the various major lobes of the brain; and can now estimate the surface area from the three-dimensional reconstruction of the cortical surface itself (fig. 5.4).

Using a statistical method that allows us to measure the amount of variance of the variations seen in twenty-seven different regions of each half brain, we discovered that the left hemisphere showed less variance in twins as compared to four regions in the right hemisphere and unrelated controls. It is interesting that females show significantly more areas in the left hemisphere. These data indicate that the development of left-hemisphere structures is under considerably more genetic control than is the right hemisphere. Given the dominance of the left hemisphere in language and in problem-solving ability, some of the cortical areas showing less variation may reflect a structural basis for the similarities in cognitive skills and personality commonly noted in twins. Also, in view of previous clinical correlations, the finding of reduced variance in the speech and language areas of the left hemisphere suggests that these systems are under greater genetic control than homologous regions in the right hemisphere.

Neuropsychological analysis of patients with focal lesions has emphasized the importance of the structures of the left hemisphere for most

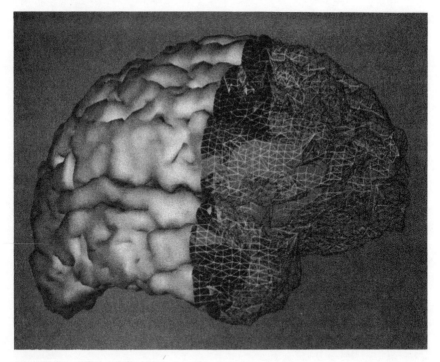

FIGURE 5.4 Three-dimensional surface model of the human brain derived from magnetic resonance images. The model, in enabling us to quantify the cortical surface area, allows us to estimate the similarities and differences between twin and nontwin brains.

measures of intelligence. For example, lesions to frontal-lobe structures can seriously disrupt cognitive processes—by modifying the capacity to switch categorical sets, verbal fluency, story comprehension, and problem solving. Furthermore, in addition to the well-known disruptions on delayed response and other tasks, lesions to frontal areas in the monkey are now thought to disrupt working memory. In this connection, it has recently been shown that variations in human working memory capacity correlate with individual variations seen in reading comprehension. The finding of reduced variance in frontal-lobe structures in the left hemisphere is consistent with the view that similar anatomical organization may reflect similar psychological capacities.

This same line of reasoning is also supported by the careful examination of split-brain patients, which has shown that the left hemisphere is far superior to the right in problem solving. These same studies have also shown that only the left hemisphere can make causal inferences and apprehend complex linguistic constructions, and thus that the left hemisphere is clearly specialized for cognitive operations in the normal brain. Related studies have also shown that performance scores on measures of intelligence are lowered following brain bisection, presumably because the neural structures associated with these activities are no longer able to contribute their processing capacities to the left hemisphere.

In summary, the present findings are consistent with the view that the previously observed similarities in psychological and physiological functions in monozygotic twins may correlate with the reduced variance seen in major cortical structures, particularly those of the left hemisphere. Those regions showing reduced variance in the left hemisphere are many of the same regions that, when lesioned, produced serious defects in intelligent behavior. Taken together, the evidence is fairly strong that specialized circuits in the left brain are managing the complex task of human intelligence.

Human Limitations

The overwhelming evidence in favor of a biological basis for the majority of intelligent behavior, combined with the fact that this capacity derived from selection processes at work over millions of years, raises the question of what we humans are *not* built to do. In the last chapter, I examined the biological roots of language and reviewed the compelling evidence for its innate structures that specify how sound is mapped onto meaning. Given that the system has been selected to work in a set way across all members of the human species, it would seem likely that there would have to be languages that humans cannot learn easily. Selection implies specificity of capacity.

Mental Specificity in Children and Chimpanzees

This idea of mental specificities is almost taken as a given when considering the intelligence of animals, even chimpanzees. Even though during the first eighteen months of human life, there are few intellectual accomplishments not paralleled in nonhuman primates, particularly the apes, they remain utterly distinct from us in their overall capacities. We can do

104

things the chimp simply cannot do. Yet both species develop object concept, imitation, spatial concepts, cause-and-effect relationships, and means-ends reasoning. Both young apes and young humans become skillful, practical experimenters. In general terms, the nature of learning processes in chimp and human infants is virtually the same. At the same time, chimps and human infants do not have identical forms of behavior. They simply have more in common in the first eighteen months than they do in later life. By adolescence, human intelligence is uniquely human, and other primate intelligence is unique to those particular species. When circuits unique to our species click in, we leave the chimp in the trees.

Several ideas concerning the mechanisms appear to make our mental life resemble that of the chimp when we are both young. It has been proposed that evolutionarily older forms of adaptation (or learning) are more likely to have limited genetic variability and a higher degree of what is called "canalization." Canalization is a genetic predisposition for a certain form of internally regulated adaptation. The genetic program does not specify a particular response to any environment, but rather specifies a generalized responsiveness to the distinctive features of environments within a certain range of variation. It has been proposed that the older parts of the brain—such as the brain stem, the midbrain, and the limbic structures—are less polymorphic than cortical areas of the brain, and contain the circuits for these kinds of behaviors. The same argument holds that the behavioral characteristics associated with higher cortical centers are newer evolutionary phenomena and likely to develop more variable phenotypes. In short, behaviors associated with older areas of the brain are genotypically and phenotypically less variable; the variation seen in human intelligence could come from variations in cortical networks.

Infant intelligence shows some signs of early canalization in the timing and the general outline of sensory-motor development. Later intellectual development, particularly around adolescence, seems to have far less definite form and timing. The evidence suggests that there is less genetic variability in early, infant intelligence than in later intelligence. Overall, the implications of animal studies are that the capacities are species-specific and limited. Can they be changed or enhanced by introducing mental aids to these animals, much as we try to introduce mental aids for our young children?

David Premack has provided evidence that certain cognitive abilities in chimps can be slightly "upgraded" in the presence of formal language instruction. While this is not to say that language instruction will enable chimpanzees to think like humans, he has shown that the ability to form

105

categories and infer causal relations is greater in those chimpanzees that have language instruction than in those that do not.

Specifically, chimpanzees (both language-trained and nontrained) were presented with tests of matching, conservation, and completion tasks. The chimpanzee with the most extensive language training, fifteen years, was Sarah. Sarah was never taught to analyze any of the actions carried out every day in her immediate surroundings. Yet, when shown an incomplete action, she completed it correctly. The other language-trained chimps did equally well. In addition, children at two and a half years of age manage these tests easily. In making correct choices, children and animals demonstrate an understanding of the basic idea of action and recognize the many different forms by which an action can be expressed. For example, when a glass of water is poured into a tall, narrow glass and a wide, short one, children below the age of five or six say that the tall glass has more water than the short one. The nonlanguage-trained chimps made similar errors; but Sarah did not. Unlike children below the age of six, Sarah conserved both liquid and solid quantity (fig. 5.5) On the other hand, it was not possible to test Sarah on conservation of number since she failed to make reliable judgments about the number of two sets of buttons. Number seems to lack salience for the chimp.

At the same time, the chimpanzees were also able to form workable mental representations of things like real fruit and actual space, yet performed surprisingly poorly when shown human representations of fruit or space. Thus, chimps have great difficulty seeing a correspondence between photos or TV images of an object and space and the actual objects and space these represent. Although chimpanzees can translate real-world objects and spaces into viable mental representations, using them in reasoning and problem solving, they are weak at doing the reverse—taking our versions of physical representations and using them as a guide to the real world.

Socially the chimpanzee differs from the human in that neither its facial expressions nor its vocalizations become voluntary: These will never be used to express intention. This contrasts markedly with young children, and correlates with a known difference between brain mechanisms that control voluntary and spontaneous facial expressions, as I shall describe in chapter 8. By age six, children can, by interpreting facial expressions, distinguish between people who really know the answer to a problem as opposed to those who are guessing. Chimps with months of training never learn the art of attribution. Yet no child needs to attend school to be formally trained in it, for, as with speech and the analysis of the word into causal relations, social attributions come naturally. Children even

FIGURE 5.5 Sarah, the famous Premack chimp, could solve problems that a five-year-old child could solve. Chimps not trained with Premack's metalanguage system could not solve such problems. Even with these skills, however, Sarah was severely limited in the other kinds of problems she could solve. Photograph by Dr. Rosemary Cogan. Used by permission.

graduate to the attribution of attribution, granting others the capacity for making attributions—a step no chimp can make no matter how much it is trained.

In the abstract code, representations can take the form of words, and words do not look at all like the things they represent. Words combine both abstract and imaginal codes. Because it has an abstract code, the chimpanzee can be taught an artificial representation system—a kind of language. However, the chimpanzee's language must be carefully distinguished from human language. In learning a language, children spontaneously observe a word order to which they adhere. Their sentences do not make a strict correspondence between words and real-world items, but are open to indefinitely large variations. The chimp's language does not have such variation. Over all, chimp language is a sorry system.

Thus, upgrading the mind of the chimpanzee by language training does not dramatically augment its cognitive abilities. Some changes in categorization and grouping are evident, but no one is claiming that formal instruction can produce a quantum leap of intelligence. Premack suggests that upgrading the mind of the chimpanzee may simulate a chapter in the history of our own species. He believes that our ancestors had the potential for representation, and that this potential underwent a substantial upgrading. The change, however, was not only in mental representation; the body itself changed, becoming increasingly able to register intention. The will of each individual could be expressed not only in movements of limbs and trunk, but in endless nuances of face and voice. At the same time the special capacities of the human brain were accumulating. The human mind became capable of attributing, to other minds around it, not only intention, but myriad other states. As I pointed out in the chapter 4, the cognitive arms race was off and running.

Languages We Can't Learn

Chimps are locked into a species with a defined set of capacities. Efforts to upgrade the chimp mind meet with limited success. We humans, however, are imperial. We think that we can do anything because we have evolved a wide range of capacities. We can absorb and learn any set of information presented to us. The fact is that we are locked in, too—as is indicated by the fact that there are languages that humans can't learn, but that other learning systems can. In a compelling analysis, Steven Pinker points out this astonishing conundrum. He writes, "Make up any finite set of sentences that includes everything that children hear (e.g., the totality of their parent's utterances, or all grammatical sentences 25 words long or less). That language is unlearnable by children, who are, nonetheless, eventually able to speak an unlimited set of sentences, limited only by memory, planning buffers, etc. It would be easily learnable by a souped-up parrot or dictation machine." For strange languages that people have actually tried to teach children, you need pretty exotic circumstances.

Consider first the phenomenon of creolization. Derek Bickerton, of the University of Hawaii, relates how the owners of colonial plantations brought together slaves and indentured servants who lacked a common language (this was done deliberately, to make it harder for them to plan subversive activities). Like many groups needing to communicate, they developed a "pidgin," a crude communication system that strung together words from the "superstrate" language (that of the plantation owners) with crude syntax and no grammatical morphemes, embedding, or other trappings of a real language. Their children, who were brought up in play groups while the parents worked in the fields, heard only the pidgin from the caretaker. Yet they didn't grow up to speak pidgin; in a single generation they created a full-fledged language, a creole with auxiliaries, tense, and embedding. Clearly, pidgins are unlearnable languages. Furthermore, Bickerton claims that historically unrelated creoles are highly similar, with a common form that points to a "language bioprogram."

As for sign languages, some sign languages in use were invented by benevolent deaf-school directors.* Here, I agree with Pinker that the language children acquire in these situations is more complex and varied than those that they are officially taught. Apparently this still happens because even children who learn a simplified, unanalyzed version of sign language from adult signers (who learned to sign after the critical period) still come up with a far more subtle grammatical system. So the invented precursors of sign language are actually unlearnable by children. In addition, as David Perlmutter, a syntactician at the University of California at San Diego, reports, children can't learn a sign language based sign-for-word on English. The reason is that languages don't allow just any old mappings between grammatical elements and their surface realizations; inflections (for example, suffixes like "-ed" and "-s") can't look just like independent words in the surface string and still be treated as part of

---

*For example, in the late eighteenth century, de l'Epée allegedly invented French Sign Language, the precursor of American Sign Language.

morphology. In this cooked-up sign language, the inflections stand as separate signs, as if they were words (for example, "I walk ed to the store"); and, sure enough, kids treat them like words, signing the equivalent of "ed I walk to the store." As for prescriptive English, no one has summed up its difficulties better than Chomsky: ". . . prescriptive English (what Fowler, William Safire, John Simon, etc., sneer at us for not speaking) is not a learnable language; if it were, they wouldn't have to keep haranguing us about it."

There are many reasons to believe that the brain does simply compute on any old input, but has circuits that specify how information will be processed. Beyond language, there are, of course, dozens of other perceptual and cognitive abilities for which we humans may also have a fixed capacity. Two psychologists who study memory, Endel Tulving and Daniel Schacter, have come across a fascinating example of this. They are experts in an experimental paradigm used in cognitive psychology called "priming," a procedure whereby stimuli are placed in an order in which a subject reacts more quickly to a second stimulus, because of the nature of the first stimulus. With language stimuli, if the word "doctor" precedes the word "nurse," "nurse" can be judged more quickly to be, say, a word. If the word preceding "nurse" is "toaster," the judgment is slower. Using this kind of facilitation data, psychologists can study the organization features of a system and, in the case of language, the organization of the lexicon.

Schacter and Tulving, while searching for principles about the organization of the memory system, came across a remarkable example of how selection appears to have shaped our brains to be prepared for processing only ecologically valid computations. In figure 5.6, the two top geometric shapes are examples of possible objects that could exist in space; the two bottom ones are impossible objects, containing properties that would violate structural principles. When normal subjects are shown these pictures, Schacter and Tulving find priming for the possible shapes but not for the impossible ones. It is as if the human brain simply, automatically, and quickly rejects a drawing that makes no sense to the cognitive system: It registers with us, but we ignore it.

Clearly then, we can do things the chimp cannot. We have specialized circuits for these abilities. At the same time, there are things we cannot do—as is nowhere more apparent than in language learning and in thinking, especially in spatial terms. Our brains are built to process things in certain ways, and no amount of education or training can take us beyond these built-in characteristics. They have been selected out and have become, over millions of years, part of our repertoire of mental

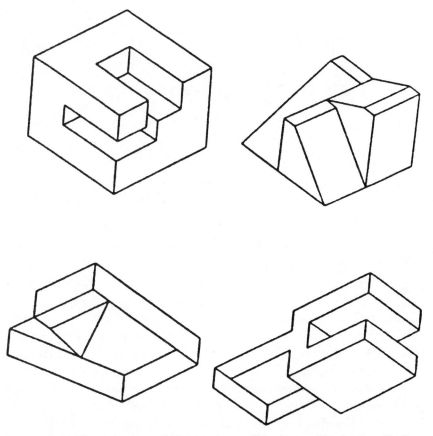

FIGURE 5.6   The Schacter and Tulving impossible figures: The brain will allow the top row to be primed but not the bottom row; the bottom row figures are impossible. Source: E. Tulving and D. L. Schacter, "Priming and Human Memory Systems," *Science* 247 (1990): pp. 301–6. Copyright 1990 by the AAAS. Used with permission.

faculties. All members of our species have the general features that allow for the basic inventiveness of the human being. It is also true that level of inventiveness varies greatly within our species. To understand that dimension of intelligence, we have to consider how the many built-in basic capacities interact to allow for human intelligence. Understanding this aspect of the problem is not an easy task. It has yet to be done.

CHAPTER 6

Selecting for Mind

I T is relatively easy to discuss selection theory with reference to brain development, perceptual processes, or even language. These are systems that work, that seem to be set properties of our species. Our perceptual system, for example, distinguishes figures from background, our language capacity insists on certain universals in syntax, and so on. As I see it, it is the human capacity to form and hold beliefs about the world and everything else that helps us adapt to much of our environment.

Beliefs are born out of the capacity to make inferences. It is one thing to register your neighbor's intention when he grimaces, and take appropriate action, but quite another to make a mental note about his character and to construct a belief about him. The next time you see him, his smiling ways may be tolerated, but your unchanged belief about his basic nature may serve you well if you should decide to turn your back. Human inferences allow you to go beyond the immediate data and generate hypotheses about something that can, in turn, deepen into beliefs.

Indeed, our overall psychological diversity demands an explanation. The *variation* that exists in each of us, the capacity to be unique, to override the rigidity of our nervous system must be addressed. Humans have a greater functional plasticity than other species, and appear to have repertoires of responses that go beyond the simple variability that allows species to adapt to changed environments through selection. Overall, animals exist in a rigid relationship with the environment. Human existence appears to be less rigid, though, and no one seems to know exactly why.

The special capacity to make an inference about both internal bodily states and external actions of ourselves and others seems, when fully developed, to reside in the left hemisphere of humans, and is called the "interpreter." The interpreter is a powerful system that is at the core of human belief formation. Without it, we would be little different from animals. With it, we become wonderfully inventive and individual even though our nervous systems are all more similar than not. The selection pressures developed in humans a capacity that saves us from being completely beholden to the environment—and thus, in some ways, has outsmarted itself.

The interpreter's central presence in the adult brain is a matter of little debate. Our species survives because of this system's activities. However, before seeing how the interpreter was discovered, it is of great interest to observe how the young child develops its inferential capacities and to watch these capacities build into the full-fledged adult interpreter.

The Infant Perceptual System

Understanding Objects

Alan Leslie, a British psychologist who believes that understanding the generation of perceptual illusions can provide significant insight into the nature of the mind, has approached the question of how infants treat causality from this perspective. Illusions occur to us even though we usually know that certain things are actually impossible, however convincingly they appear—as when driving across the Nevada desert, you see a mirage. Leslie feels that what he calls the "incorrigible presence" of such a mirage must be due to some limitation in the brain with respect to the way it builds up the raw sensory data from the eyes to create a perception.

Leslie's idea is that there can be a perceptual illusion of causality. This implies both that a perceptual mechanism operates automatically on our perception of events, producing abstract descriptions of their causal structure; and also that the idea of cause and effect does not originate in prolonged learning. Indeed, infants do seem to know that solid objects cannot inhabit the same space at the same time, even temporarily—the principle of "no cohabitation." For example, in Baillargeon's experiment reported in chapter 3, five-month-old infants were habituated to a screen that rotated up and down on a table like a drawbridge. Once the infants were habituated to this motion, a box was placed behind the screen. The infants then watched the screen go through the same motion in a possible and an impossible condition. Baillargeon noted that the infants who were shown the screen moving through the box, an impossible situation, evinced greater interest and appeared to be surprised. The infants who saw the screen stop when it reached the box showed less interest and were not surprised. This experiment indicated that infants have a representation of objects, which they are able to use to generate expectations about how an object ought to behave. Indeed, infants are born with the processes that automatically allow for this kind of inference already built into the brain. The perceptual problem has been framed by capacities with which the infant brain is innately equipped.

The adult visual system also has a representative understanding of objects, yet does not always act in accordance with what it knows. This system is prepared to accept even bizarre perceptions in the presence of a sufficient illusion. For example, in the illusion of the Pulfrich double pendulum (PDP), one views a simple pendulum with a medium-density filter over one eye (fig. 6.1). The motion of the pendulum creates the illusion of an elliptical path. Adults consistently report seeing this illusion, even though they know that such an event is not possible. Thus, we have the ability to detect the inconsistencies, but lack the ability to modify what we see.

Leslie maintains that this dichotomy would not be possible if the visual input systems were simply mechanisms of associative learning or instruction, and suggests that a central system must account for the principle of no cohabitation. This system maintains a consistent model of the infant's (and adult's) current environmental situation. Visual input is perceived in terms of this central mechanism, which may well have another related function of building encyclopedic knowledge and commonsense theories about the world.

Leslie concludes:

The main organizational features of the adult mind appear to be present in infancy. Central thought processes appear to operate early and, like perceptual processes, are richly structured, presumably by biological endowment. They employ powerful inferential processes which are sensitive to the logical properties of infant symbolic representation. Towards the end of infancy, thought acquires the power to represent itself recursively and thereby to reason imaginatively. This distinction allows central processes to theorize about those things in experience that are incorrigibly not what they seem. The main

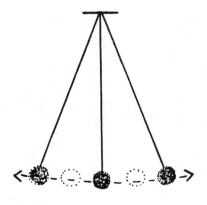

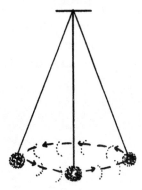

FIGURE 6.1 The Pulfrich double pendulum experiment: The upper figure is the path of the pendulum; the lower, that path as seen when one eye is covered by a medium density filter.

conclusion appears to be that human mental architecture provides the basis for development and not its outcome. Acquiring theoretical knowledge of the world—in the sense of both common sense and of more specialized scientific and religious theories—is uniquely the point of human development. The basic organization of the infant-adult mind is highly designed for this task.

(Leslie, 1988, pp. 207–8)

The idea of a central mechanism evolving to allow for these crucial human developmental events has also been examined for the past ten years by Elizabeth Spelke of Cornell University. She suggests that infants do not perceive the unity, boundaries, and persistence of objects under all the conditions that are effective for adults. Rather, she posits that the mechanism by which infants perceive the world is central, "so central that it may be misleading to say that objects are perceived. Objects may be known instead by virtue of an early developing theory of the physical world."

Infant studies provide evidence that the mechanism by which infants apprehend objects is more central than traditional theories had suspected. The mechanism of object perception appears to analyze properties as they are perceived, as opposed to operating on individual patterns of stimulation. The mechanism also seems to be amodal, accepting input from different perceptual systems; and to enable infants to go beyond the world of immediate perception, allowing them to make sense of events in which objects are hidden and predict the future behavior of those objects.

Evidence that the mechanism for apprehending objects operates on a representational level comes from an experiment by Spelke and her colleagues (fig. 6.2). In this experiment, infants' perception of partly

FIGURE 6.2 The experiment that Elizabeth Spelke and her colleagues devised to test infants' perception of objects in the real world.

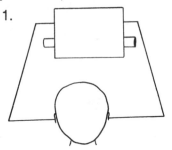

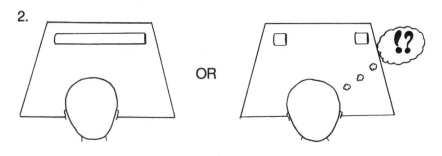

hidden objects was examined by asking whether infants, below the age of being able to carry out visually guided reaching, perceive an object (such as a rod with a box in front of it) as two separate images, or as one object (the rod) occluded by the box. Four-month-old infants were presented with an image of a rod occluded by a box until they were habituated to the image. The infants were then presented with an image of the rod as a whole object or as two fragments, with a gap where the box occluded the object. Infants showed more surprise at the broken fragments, indicating that they perceived the rod as a whole unit and not as the sum of its visual components. Young infants could detect the properties of the display, but did not use these perceived properties when organizing the perceptual field into objects.

To examine the role of different modes of perception, Spelke focused a set of experiments on object perception to determine whether infants perceive the unity and boundaries of objects under the same conditions when they feel surfaces as when they see them. In this study, four-month-old infants held two rings, one in each hand, under a cloth that blocked their view of the rings and their own bodies. In the first experiment, the rings could be moved either independently or only rigidly together. Infants were allowed to move the rings at will. To investigate whether infants perceived the independently movable rings as separate objects and the rigidly movable rings as connected, half of the infants were habituated to a tactile display and then shown alternating visual displays of connected and separated rings undergoing no distinctive motion.

In these experiments, habituation to the independently movable rings was followed by greater generalization to the separated display, providing evidence that infants perceived the independently moving rings as distinct objects. In contrast, habituation to the rigidly movable rings was followed by greater generalization to the connected display, providing evidence that infants perceived the commonly moving rings as a single object. These findings provide evidence that objects are perceived under the same conditions whether they are seen or felt.

Spelke concludes that the principles of cohesion, boundedness, substance, and spatial-temporal continuity are central to thought for our species in childhood as well as in adulthood. The core of knowledge that is presumably inherited is built upon throughout life and results in what amounts to every adult's ability to perfectly predict a variety of physical events, such as the manner in which paths connect through unoccupied space. Spelke predicts that a child who does not come equipped with a core or initial theory will not develop a systematic theory about the knowledge in question.

Developing a Theory of Mind

Much work has also been carried out on how and when the child develops a theory of mind—rational processes that lead to the formation of beliefs. When discussing this issue, cognitive scientists always refer to "representational systems," meaning that once information has been presented to a brain, it is constantly being modified, or represented in another form. Thus, if the sentence "The girl was kissed by the boy" is read to a subject, it will take that person a certain amount of time to judge which of four subsequent sentences are consistent with the statement. In this example, which was invented many years ago by the Stanford psychologists John Anderson and Gordon Bower, the four sentences were:

1. The boy kissed the girl.
2. The girl kissed the boy.
3. The boy was kissed by the girl.
4. The girl was kissed by the boy.

The fact that sentence 4 was responded to more quickly than sentence 1 led these researchers to believe that the mind immediately preserves the active/passive distinction. When, however, the subjects were tested two minutes later instead of immediately, they responded more quickly to sentence 1, thereby suggesting the information had been transformed (represented) into another, active form.

Psychologists have been puzzling over how such representational systems emerge in the child's cognitive life. Here again, as with perceptual phenomena, the child appears to start with considerable built-in capacity. It is generally agreed that the two-year-old child is, however adorable, not yet working with a full deck. As every parent knows, simple desires are the rule, and the child moves everything in its path to obtain them. While the cortex that overlays the more primitive subcortex is functioning—as we know, since the child can understand language as well as speak—the intricate cortical circuits that subserve representational systems are not yet all functioning. As a child develops, so does the brain.

At the age of three, the child's mind begins to explode. Suddenly the child has a mini-version of the adult mind. According to recent studies by Henry Wellman at the University of Michigan, the three-year-old child can sort out the implications of stories that involve manipulating what the characters believe and how those beliefs have consequences for the story characters. Moreover, in stories having what he calls a "belief-desire conflict," three-year-olds can see how a character's beliefs can override desire.

Although Josef Perner at Sussex University feels that these capacities come a little later, both Wellman and Perner see the importance of the phenomenon of misrepresentation of information. Children around the age of four or five start to engage in deception, or the misrepresentation of information to others. Here again, an entire way of interacting socially pops into the child's consciousness, as if a switch had been turned on. Prior to the age of four, games of deception go nowhere. In one recent experiment by Joan Peskin children were told that two puppets would approach them and ask which of four stickers was their favorite. It was known that each child had a favorite sticker. They were also told that one particular puppet was a friendly sort and, after asking, would choose another sticker. The other puppet, however, was beastly and would most certainly grab the child's favorite. The three-year-olds were unable to deceive, while, from age four on, children quickly learned to trick the mean puppet while remaining truthful to the nice one. It seems unarguable that tremendous selection pressure is at work in the species to develop such strategies.

The evidence that the brain develops in a way that makes sense with the cognitive findings has always been a part of neurobiology. It has been maintained for years that the human cerebral cortex is slow to develop (fig. 6.3). In particular, the sheaths that wrap around neurons in order to allow them to function efficiently, a substance called "myelin," is slow to be established in the brain. Once there, it appears to last for the brain areas known to be associated with higher cognitive functions. In addition, it is well known that synaptic processes continue to be established late in human development. Such observations are consistent with the view that circuits are coming "on-line" through early development, circuits that could well serve new cognitive capacities.

All of this suggests that the interpreter is an emerging capacity, whose development can be traced from infancy. It allows the child to develop a theory of mind and the capacity to distinguish itself from others. Its crucial role in calculating the intentions of others finds it to be a system that evolutionary pressures would have selected out in an efficient way. In a world of limited resources and other dangers, organisms with such a capacity would clearly be at an advantage. It also suggests that if a human had a dysfunctional interpreter, he or she would be a very disordered human being. Henry Wellman has suggested that such a state of affairs exists with the autistic child, who cannot function in a social world. At the extreme, such children cannot have commerce with any adult, and

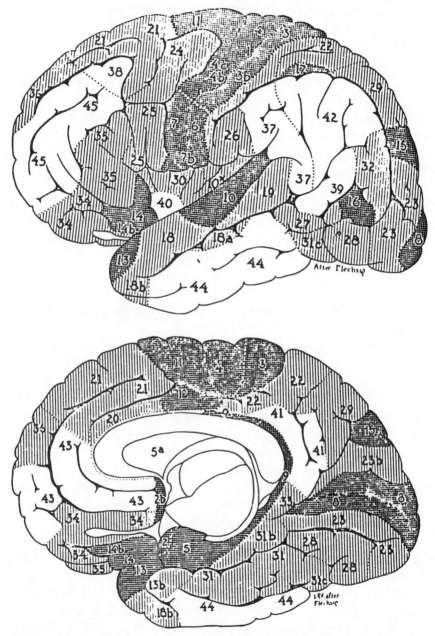

FIGURE 6.3 Myelinization patterns of the developing human brain. The clear areas are both the last to develop and crucial for higher cognitive function. Reprinted with permission from Percival Bailey and Gerhardt von Bonin, *The Isocortex of Man* (Urbana: University of Illinois Press, 1951), p. 7.

adults are at a loss when explaining to them the nature of things. On the other hand, the retarded child is totally different. Full knowledge of the pathologic correlates of these disease types may provide clues to the exact brain networks involved in the interpretive function.

THE ADULT INTERPRETER

The discovery of the interpreter came slowly over the years, a fruitful product of examining the nature of conscious mechanisms in split-brain patients. These studies revealed that the brain is organized in terms of modules. The idea of modularity is central to modern cognitive neuroscience and will be reviewed below. It is also worth describing how the interpreter mechanism was discovered, as the history of the issue also reveals some of the limitations of early claims that splitting the brain would produce two sets of conscious controls. As the work has evolved, we now see that only the left hemisphere has the underlying brain organization which allows for high-level consciousness and belief formation.

In the fall of 1961 I sat in an isolated room at the California Institute of Technology, examining a human who had undergone brain bisection, known as case W. J. The testing procedures were crude by modern standards, but the idea was clear. A war veteran had had his two half brains disconnected in an effort to control his epilepsy. Although it was a radical procedure by anybody's standards, a flurry of animal experimentation around that time had encouraged the idea that bisection of a human brain would produce dramatic results.

The antecedent animal work on cats and monkeys had sectioned the nerve fibers connecting the two half brains and the eyes, and had shown them to be bewilderingly unique. Information that had been trained to one half brain was not known to the other. It was as if, literally, the left half of the brain did not know what the right was doing. The likelihood of this simple idea being true for a human being was considered remote, which may explain why I, a new graduate student, was assigned to the task of examining the first patient. Weeks before W. J.'s hospitalization, Joseph Bogen, the neurosurgeon who assisted with the operation, and I had examined the patient in detail. With his brain not yet operated upon, he behaved like any normal person. If information was lateralized to one half brain, the other knew of it instantaneously. The great cerebral commissure connecting the two half brains worked and kept each half brain up to date with what was known by the other. Or, so the story went.

Also, months before the first case (W. J.) appeared, another patient with a partial section had been tested by a research fellow, with no spectacular results. The profile was consistent with an earlier history of commissure section in humans; and the general idea in the lab was that humans were somehow different from animals. All of these views changed in one afternoon.

After his split-brain surgery and like all subsequent relevant cases, case W. J. normally named and described information presented to his right visual field (that is, when information was made available to his left hemisphere). We were surprised by his apparent lack of response to stimuli presented to his left visual field, which, of course, were available only to his surgically isolated right hemisphere. It was as if he were blind to subjective visual stimuli presented to the left of fixation—for which possibility we were prepared and I had designed the tests to allow for a manual response. But, almost immediately, it was obvious that while the left talking hemisphere could not report on these stimuli, the right brain, with its ability to control the manual responses of the left hand, could easily react to the simple visual stimulus. In that dramatic afternoon the modern era of split-brain research was born.

We made our first evaluation of the data that started to pour in during those early years. After the hemispheres were divided to control intractable epilepsy, each half brain behaved independently of the other. Information experienced by one side seemed unavailable to the other. Moreover, each half brain seemed specialized for particular kinds of mental activity. The left was superior for language, while the right seemed more able to carry out a variety of visual-spatial tasks. We had separated structures with specific functions. We talked easily about separated consciousness and did so without defining our terms. We did that by design, since no one seemed to know exactly how to define human consciousness. We simply listed capabilities of each side of the brain and rhetorically asked whether anyone, considering this data, wouldn't argue the same.

The capacities possible from the left hemisphere were no surprise. However, when we discovered that some patients were able both to read from the right hemisphere and to take that information and choose between alternatives, the case for a double conscious system seemed promising. We could even provoke the right hemisphere in the sphere of emotions. Over a period of five years, I carried out dozens of studies on these patients, and all results pointed one way: After separating the human cerebral hemispheres, each half brain seemed to work and function outside the conscious realm of the other; each could independently learn, remember, emote, and carry out planned activities.

This set of data was reported in the early 1960s. It was suggested at the time that there was evidence in hand that opened up the difficult question of whether each of the separated hemispheres has its own free will, its own conscious reality.

Time, more patients, and more studies modified the original picture. New observations in the early 1970s not only challenged the simple view of hemispheric functioning and ideas about double consciousness, but also offered a new conceptual framework, which threatened still further the existing concepts about the unity of conscious experience and the belief that all human action is generated by a single rational process. The studies we carried out on a series of new patients operated on in the eastern United States confirmed the broad outlines of the original California series, spelling out the basic properties of each hemisphere and the dramatic disconnection effects. In fact, three of our patients were more advanced in right-hemisphere capacity than the two earlier California patients, cases N. G. and L. B. Two of the three cases—P. S. and V. P.—were able both to speak and to comprehend simple language in their right hemisphere.

The capacity of these patients, however, stood in marked contrast to other new patients who possessed little or no right-hemisphere language. These patients were also very poor at carrying out other perceptual and cognitive tasks. This contrast led us to the view that language is central and critical for cognitive functioning in the right hemisphere. Also, at this time we discovered perhaps the most important aspect of the dominant left hemisphere, a capacity that may go a long way toward defining the difference between human beings and animals. The functioning of this special left-brain system, the "interpreter," is best understood in light of new views on how the human brain is organized.

Until the mid-1970s, the idea was that each half brain had the full biological complement of "stuff" that allowed for the essentials of human conscious experience. While language and speech gave the left hemisphere the edge, its superiority was thought to be more quantitative than qualitative, and the left hemisphere was not yet seen as possessing some special capacity other than language and speech.

For years, the big units of psychological life—such as visual or auditory perception, semantics and syntax, or attention—had been dealt with as whole units in brain science, not as final products of interacting and interconnecting subprocesses, as if a vast yet unified system in the brain produced our personal conscious reality. It was assumed that splitting the brain revealed secrets about this system in that it had been duplicated by Mother Nature and could exist independently in one hemisphere or the other if the two were divided.

At about this time, all sorts of cognitive phenomena were beginning to be broken down into their component processes. New discoveries suggested that the brain is indeed organized in a modular fashion with multiple subsystems active at all levels of the nervous system and each processing data outside the realm of conscious awareness (fig. 6.4). These modular systems are fully capable of producing behaviors, mood changes, and cognitive activity. This activity is monitored and synthesized by the special system in the left hemisphere, the interpreter. The right hemisphere does not have such a system, since it does not have other aspects of a logical-deductive system. In short, the new studies showed the error of the idea of a doubling consciousness: While many basic functions are bilaterally represented, those essential for human thought are not.

The modular organization of the human brain is now fairly well accepted. The functioning modules do have some physical instantiation, but the brain sciences are not yet able to specify the nature of the actual neural networks involved for most of them. It is clear that they operate largely outside the realm of awareness and announce their computational products to various executive systems that produce behavior or cognitive states. Catching up with all of this parallel and constant activity seems to be a function of the left hemisphere's interpreter module. The interpreter works under strict experimental conditions in dramatic ways.

We first revealed the phenomenon in the early 1970s using a simultaneous concept test. In this task, the patient is shown two pictures, one exclusively to the left hemisphere and one exclusively to the right, and asked to choose from an array of pictures placed in full view in front of him the ones associated with the pictures lateralized to the left and right brain. In one example of this kind of test, a picture of a chicken claw was flashed to the left hemisphere, and a picture of a snow scene to the right hemisphere. Of the array of pictures placed in front of the subject, the obviously correct association is a chicken for the chicken claw and a shovel for the snow scene. P. S. responded by choosing the shovel with the left hand and the chicken with the right: That is, the right hemisphere matched the snow scene. When asked why he chose these items, his left hemisphere replied, "Oh, that's simple. The chicken claw goes with the chicken, and you need a shovel to clean out the chicken shed." Here, the left brain, observing the left hand's response, interpreted that response

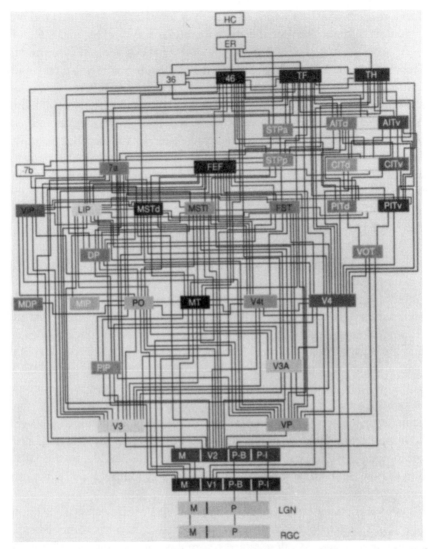

FIGURE 6.4 David Van Essen has best described the modular organization of the visual system of the primate. Each year he seems to unearth more and more subcomponents of our visual system, leading to intricate cortical maps which look more like a computer-chip-wiring diagram. The various letters refer to specific subregions of the visual system. Reprinted from Daniel J. Felleman and David Van Essen, "Distributed Hierarchical Processing of the Primate Cerebral Cortex," *Cerebral Cortex* 1 (1 [1990]): 30, with the approval of Oxford University Press.

into a context consistent with its sphere of knowledge—one that did not include information about the right hemifield snow scene (fig. 6.5).

Another example of this phenomenon of the left brain interpreting actions produced by the disconnected right brain involves lateralizing a written command, such as "laugh," to the right hemisphere by quickly presenting it to the left visual field. After the stimulus was presented, one patient laughed and, when asked why, said: "You guys come up and test us every month. What a way to make a living!" In still another example, when the command "walk" is flashed to the right hemisphere, patients will typically stand up from their chairs and begin to leave the testing van. When asked where he or she is going, the person's left brain says, for example, "I'm going into the house to get a Coke." However you manipulate this type of test, it always yields the same kind of result (fig. 6.6).

There are many ways to influence the left-brain interpreter. As already mentioned, we wanted to know whether the emotional response to stimuli presented to one half brain would have an effect on the affective emotional tone of the other. In this study, under lateralized stimulus-presentation procedures, we showed a series of film vignettes that included either violent or calm sequences, and used an eyetracking device that permits prolonged lateralization of visual stimuli while the eyes remain fixated on a point. The computer-based system keeps careful track of the position of the eyes so that if they move from fixation, the movie sequence is electronically turned off. For example, in one test a film depicting one person throwing another into a fire was shown to the right hemisphere of patient V. P. She reacted, "I don't really know what I saw. I think just a white flash. Maybe some trees, red trees like in the fall. I don't know why, but I feel kind of scared. I feel jumpy. I don't like this room, or maybe it's you getting me nervous."

As an aside to a colleague, she then said out of my earshot, "I know I like Dr. Gazzaniga, but right now I'm scared of him for some reason." Clearly, the emotional valence of the stimulus had crossed over from the right to the left hemisphere. The left hemisphere remained unaware of the content that produced the emotional change, but experienced and had to deal with the emotion as it interpreted it.

The same kind of phenomenon is observed when more neutral stimuli are presented to the right or the left hemisphere, such as scenes of ocean surf, nature walks, and the like: The patient becomes calm and serene. Taken together, these examples show that both covert as well as overt responses are interpreted.

While patients possess at least some understanding of their surgery, they never say things like, "Well, I chose this because I have a split-brain,

FIGURE 6.5 The top frame shows one of the initial stimuli sets and how the patient, talking out of the left brain, responded to the examiner's query about his choice. Reprinted from Michael S. Gazzaniga and J. LeDoux, *The Integrated Mind* (New York: Plenum Press, 1978), fig. 5.1, p. 147, with permission.

FIGURE 6.6 When the silent right hemisphere is given a command, it carries it out. At the same time the left doesn't really know why it does so, but it makes up a theory quickly. Reprinted from Michael S. Gazzaniga and J. LeDoux, *The Integrated Mind* (New York: Plenum Press, 1978), fig. 5.2, p. 149, with permission.

and the information went to the right, nonverbal hemisphere." Even patients who, based on IQ testing, have higher IQs than P. S., view their responses as behavior emanating from their own volitional selves and, as a result, incorporate this behavior into a theory to explain why they behave as they do. Certainly, at some future point a patient may choose not to interpret such behaviors because of an overlying psychological structure that prevented the response. Or, a patient who has learned by rote, as it were, what a "split brain" is all about, and why, therefore, a certain behavior most likely occurs, might well not offer an explanation.

Some patients who have trouble controlling the left arm due to a transient state of dyspraxia tend to write off anything the arm does under the direction of the right brain, and thereby make the foregoing test inappropriate for demonstrating the phenomenon. In such situations, a single set of pictures is presented, and only one hand is allowed to make the response. For example, in this test the word "pink" is flashed to the right hemisphere, and the word "bottle" to the left. Placed in front of the patient are the pictures of at least ten bottles of different color and shape. When this test was run on J. W., on a particular day when he kept saying his left hand did what it wanted to do, he immediately pointed to the pink bottle with his right hand. When asked why he had done this, J. W. said, "Pink is a nice color."

The effects seen in these special patients and under these laboratory conditions can be related to many everyday experiences. Consider how often we go to bed in a good frame of mind (or the opposite) only to awake feeling depressed and cranky (or the opposite). If the cognitive data structure—which is to say, the facts about one's life—hasn't changed during the night, why the change in mood? Can it be that prior memories have become activated and unleashed biochemical mechanisms that give rise to a specific mood state? The idea here is that the left brain interpreter would try to make sense out of these feelings and may well (albeit somewhat gratuitously) attribute a cause for them to otherwise innocent concepts also existing in the conscious realm at that time.

The Properties of the Interpreter

Certainly, language is the vehicle through which the interpreter expresses its actions. But, as we came to appreciate the power of the interpreter and its central role in human behavior, we began a series of studies to examine whether it is also inexorably linked to language mechanisms per se. Did the fact that the right hemisphere becomes more responsive when language is incorporated into its structure provide a clue to the question?

In the patients we had studied so far, the interpreter had been represented in the left hemisphere and not in the right. In trying to understand this phenomenon, we examined more closely the language structures of the right brain in these special patients as well as whether these patients could carry out out simple problem-solving tasks.

As already mentioned, split-brain patients without right-hemisphere language have a limited capacity for responsiveness to patterned stimuli, ranging from none at all to the capacity to make simple matching judgments. The patients who had the capacity to make perceptual judgments not involving language had no ability to make a simple same/different judgment within the right brain when both the sample and the match were lateralized simultaneously. In other words, when a judgment of sameness was required of two simultaneously presented figures, the right hemisphere failed. This profile is commonly seen in patients of all kinds, including ones of similar and sometimes greater overall intelligence than those who possess some right-hemisphere language.

This minimal profile of capacity stands in marked contrast to the patients with right-hemisphere language. The right brain of these patients is responsive, and their overall capacity to respond both to language and to nonlanguage stimuli has been well catalogued and reported. Among the East Coast patients, J. W. understood language and had a rich right-brain lexicon as assessed by the Peabody Picture Vocabulary Test as well as other special tests. At the same time, J. W. had no capacity to speak out of the right brain. Studies on cases V. P. and P. S. revealed these patients were able both to understand language and to speak from each half brain. Would this extra skill lend a greater capacity to the right brain's ability to think or interpret the events of the world?

The picture that emerges is basically that, while the right hemisphere can possess an extensive auditory and visual lexicon and sometimes even speak, it is severely limited in its capacity to use syntactic information in comprehension. Thus, the right hemisphere has difficulty understanding semantically reversible active and passive sentences and recognizing the differences between phrases like "the flying planes" and "flying the planes." Recently, however, we have shown that these same right hemispheres are capable of judging whether a spoken sentence is grammatical, which means that the right brain can be critical without being knowledgeable. It can judge the grammaticality of an utterance, but it cannot use syntactic information to place constraints on understanding word strings.

Nonetheless, the right hemispheres of both patient groups turn out to be poor at making simple inferences (fig. 6.7). For example, when shown two pictures, such as a picture of a match and then a picture of a wood pile, the right hemisphere cannot abstract their causal relation and choose the proper result—that is, a picture of a burning woodpile. In other testing, simple words were serially presented to the right brain, the task being to infer the causal relation between the two lexical elements and pick the answer from a list of six possible answers printed and in full view of the subject. A typical trial would consist of words like "pin" and "finger" being flashed to the right brain, with the correct answer being "bleed." While the right hemisphere could always find a close lexical associate of all of the words used when tested separately, it could not make the inference that "pin" and "finger" should result in the answer "bleed." The successful completion of a task under these conditions must mean that the left hemisphere was controlling the response.

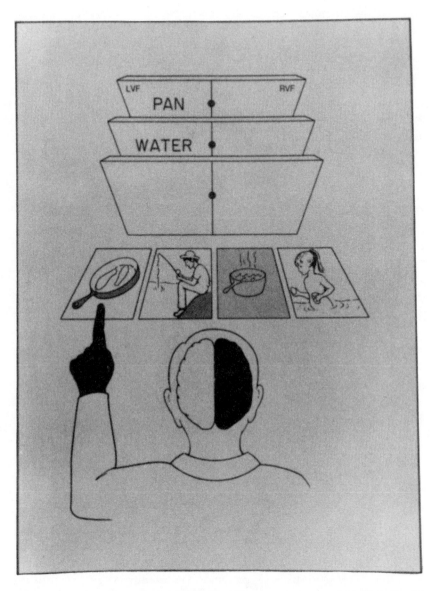

FIGURE 6.7 Even language-rich right half brains are poor at thinking. In this simple task, all that is required is that each half brain see the relationship between the two pictures and choose which of four are most related. The left brain finds these tasks as simple as a two-year-old would find them, while the right stumbles along and makes many errors.

Still, the right hemisphere was free to inspect and watch how the task was done.

In V. P., the tests were pursued and further simplified. Instead of two words being flashed to either the left or the right visual field, one word was spoken, and the other was then lateralized to either the left or the right brain. For example, the word "pin" was spoken, and then the word "finger" was flashed. This simplification did not seem to make a difference. The right hemisphere remained poor at carrying out the task.

These tests suggest that language is called upon to label and express the computations of other cognitive systems. The simple language systems of these right hemispheres are not able, in and of themselves, to perform cognitive activities. The systems that do generate cognitively based inferences are not present in the disconnected right hemisphere. There seems to be a specific brain system committed to making our species the "believing" species. We still have no evidence proving that the interpreter is present in the right hemisphere. A half brain might possess an interpreter, but because it usually doesn't talk, it can't report its interpretations. One way to examine the possibility of the interpreter's presence in the right hemisphere is to allow for a drawing, as opposed to a spoken response. In one test we asked each half brain to draw what it saw, and to each one we flashed, at the midline boundary of the visual field, a picture of half a person or a house. In this test, the left hemisphere immediately assumed that the rest of the house had also been flashed, but had fallen into the bad visual field; as a consequence, it drew us whole figures. The right hemisphere, which had the same stimulus setting, made no interpretations and drew only half figures (fig. 6.8).

Another recent study on memory suggests that only the left hemisphere interprets the information it receives. Elizabeth Phelps and I showed to each hemisphere a story book of some forty events that dealt with a man getting up in the morning and going to work. Two hours later, we tested each hemisphere separately by presenting pictures from the original story, pictures that were not presented but were related to the original story, or pictures that had nothing to do with the original story. The left hemisphere responses were similar to those of normal, whole-brain behavior in that, in addition to recognizing the actual story line, they also included possible things that could have happened. This was not the case with the right brain, which tended to stick to the original story line, making no interpretations.

LVF **RVF**

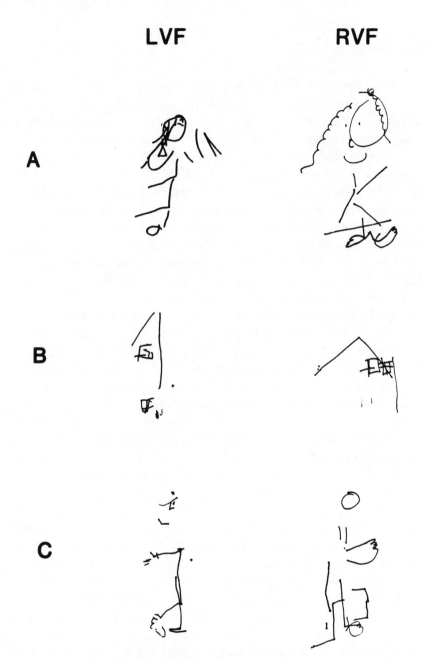

A

B

C

FIGURE 6.8 Drawing of each half brain in response to a lateralized stick drawing of half a person or half a house. The left brain interprets the stimulus and tends to complete it while the right returns exactly what it saw.

Beliefs and Behavior

It is easy to imagine selection pressures promoting an interpreter mechanism in the human brain. A system that allows for thought about the implications of actions, generated both by others as well as the self, will grasp a social context and its meaning for personal survival. Thus, you can both carry around and have access to your own "video camera" of events in which you are continually involved, and think about the impact of your own actions on your working environment. Furthermore, you can come to see the difference between your public and private selves, as you realize that others think of you in terms of their interpretations of your actions. You come to learn that the public's theory of who you are can be different from your own personal theory of both who you are and how you feel. Human beings learn quickly that feeding these two different selves is a major function of human existence and survival.

Also, the interpreter system generates the possibility of human uniqueness. Selection theory is hard on the nature/nurture issue in arguing that all we are doing in life is catching up with what our brain already knows. We are discovering built-in capacities. While this idea can appear dismal and depressing, I think that the built-in capacity of the interpreter gives each of us our local and personal color. After all, it works by drawing upon the unique experience each of us possesses. On average, the same interpretive system working in two different cultural contexts will come to different conclusions about what is important in life. How deeply it considers its experience will depend on the limits of its other operating systems, such as working memory capacity and so on.

Studies on the interpreter suggest that we humans base many of our inferences about personal and environmental events on what is happening in our immediate state of consciousness. We make use of the data we can call up on the spot, and this plays a powerful role in determining the nature of the belief or explanation we develop for events in our lives. Scientists directly studying the nature of human beliefs have made related observations. People who are provoked to do and say things in slightly different ways come to believe quite different things about their behavior. Slight, transient events in our consciousness can have powerful, long-ranging effects on what we think—even for brand-new beliefs, and even if the beliefs are subtly influenced or induced by an experimenter.

For example, a recent study asked people to provide reasons for a particular set of behaviors by completing a lead-in sentence. Some subjects were asked to complete the sentence "I engage in this behavior because——," while others were asked to complete the sentence, "I engage in this behavior in order to——." The investigators found that the word "because" tended to elicit more intrinsically motivated reasons from the subjects, while "in order to" tended to elicit extrinsically motivated reasons. The different causal explanations elicited by the experimenters influenced the subjects' later perceptions of the desirability of both the behavior and the goals it achieved.

Thus, the interpreter has a set of properties. Clearly, early information is crucial to the development and maintenance of beliefs, as is the primacy effect, whereby information presented early has a disproportionate influence on final judgment. Lee Ross and Richard Nisbett, two psychologists who have studied beliefs in detail, have recently sought to explain the tendency to rely more heavily on one set of information over another in terms of certain biases people use to create beliefs. These biases focus on the availability of information when one makes causal inferences, and on those factors that might affect the availability of some types of information. As already seen in the studies of the interpreter, beliefs seem to be built up from people's immediate experience.

How available information is for interpretation may be based on the frequency and likelihood of particular events being accessible from memory. The fact, however, that many factors can influence an event's salience can make available information misleading. Some years ago, for example, Amos Tversky and Danniel Kahneman asked subjects to judge whether words beginning with the letter R or K occur more frequently in English than words in which the letter R or K appears as the third letter. Subjects consistently and erroneously estimated that there are more words beginning with R or K. Tversky and Kahneman suggested that subjects may believe that there are more words beginning with these letters because it is easier to generate a list of words by first letter than by third letter. The belief about word frequency in this case was influenced by how well the words were retrieved from memory. On a larger scale, this finding suggests that the tendency to find confirming evidence for a belief may be a result of the fact that it is easier to think of evidence *for* something than *against* it.

Other factors that seem to influence people's ability to infer causality are the way in which a given cause resembles the effect to be explained, and the fact that a given event can have only one sufficient cause. Nisbett and Ross have pointed out, "People have strong a priori notions of the types of causes that should be linked to particular effects, and the resemblance criterion often figures heavily in such notions." The resemblance criterion is obvious in the magical thinking of certain cultures. In medicine, for example, curative agents are sometimes thought to resemble either the properties of a disease or properties opposite those of the disease. For example, in some cultures epilepsy is thought to be cured by a drug made from a monkey whose movements appear epileptic. Even though the relationship between cause and effect is not real, the link is more available because of the tendency to assume that cause and effect resemble one another.

The human capacity to hang on to our beliefs in the presence of confounding data is astounding. In one study, subjects were presented with the results of two supposedly authentic studies on the deterrent effects of capital punishment. The subjects had previously indicated whether they strongly believed or did not believe in capital punishment as an effective deterrent. The design was constructed so that one group of subjects read about the results and methods of an empirical study supporting their own position, and then read about the results and method of a study opposing it. A second group of subjects read about the opposing study first, and then a study supporting their view. It was found that subjects' belief in their initial position was strengthened if the first study supported it, while their initial belief was hardly affected at all if the study opposed it. Clearly, people evaluate evidence concerning an established belief in such a way as to maintain its perceived validity, and are more likely to discredit methodology than their own beliefs.

We human beings, with our powerful tendency to create and maintain beliefs, readily generate causal explanations of events and actively seek out, recall, and interpret evidence in a manner that sustains our personal beliefs. This tendency, which Nisbett and Ross have given the label, "belief perseverance," has been shown to be true even under circumstances in which such causal explanations have no empirical foundation. Together with some older work in the literature on impression formation, the work of these investigators suggests that people tend to view assymmetrically evidence regarding an established belief. In other words, we place a disproportionate amount of credibility on evidence that supports an established theory and tend to discredit opposing evidence. When new beliefs are formed, we tend to rely most heavily on the initial set of information used to establish the theory. Of course, the new theory, once formed, becomes resistant to change in the face of subsequent evidence.

To sum up, it appears that selective processes have, over millions of years, identified brain networks that endow new members of the human species with crucial information about a variety of matters. Studies on basic perceptual processes reveal that the young mind comes fully equipped to deal with the nature of the physical world. The automatic inferences made about the sensory world are surely useful for many species. This early and fundamental system, in providing principles of perception that are difficult to override, seems to set the stage for the higher-order inferential mechanisms that appear unique to our species. When the interpreter goes to work on more complex events, the resulting hypotheses and beliefs about the world also seem resistant to change. Even though the similarities are striking, the quintessential human property of mind—rational processes—can occasionally override our more primitive beliefs. It isn't easy, but when it occurs, it represents our finest achievement.

CHAPTER 7

Addictions, Compulsions, and Selection Theory

I MAGINE the emergency room at Bellevue Hospital in New York City on a hot summer Saturday night, when the place is jumping with the raw problems of city living. People are stuck in hallways, some dying of AIDS, some with strokes, some with broken limbs. Suddenly in walks yet another teenager, this one literally out of his mind from drugs. Enraged about something and loaded with alcohol and PCP, the young buck jumps around all the medically ill like an overactive ape. The staff is already overstressed. The drug addict steals the show, and all those there to help are furious that society somehow allows this to happen. Disease is disease, but taking drugs is voluntary administration of destructive agents. This activity must be stopped. Our species did not evolve for this purpose. Stop drugs!

The overall picture is that a lot of compulsive behavior can be viewed as having a biological component. Many addictions could be manifestations of a biological factor or of other disorders that encourage the use of substances or experiences that help relieve a particular psychological state. Looking for purely psychological explanations of why people become impaired by an addiction or a compulsion seems misplaced. It appears that people with certain complex biological substrates are selected out and fall prey to certain environmental contexts.

This common scenario makes for good television, with concomitant outrage from the public. My little Billy must be protected from drugs. Drugs of addiction are the strongest case for the argument that environment can affect and ruin lives. The assumption is that, once exposed to such substances, a perfectly good person can go down the tubes. The environment is all powerful and instruction is the rule. On this issue, selection theory has nothing to offer—or such is the prevailing attitude with respect to this and other maladaptive behavior.

When dealing with the issue of maladaptive behaviors, evolutionists divide into at least two camps. The sociobiologists, those who believe that all human behavior, including today's, is Darwinianly adaptive, have to explain things like drug addiction, homosexuality, and so on, as they suffer with the assumption that adaptation must be what is called "inclusive fitness-maximizing." Everything an organism does must be explained in terms of genetics. This popular approach seems wrongheaded to me.

Selection theorists, who, on the other hand, trumpet the evolutionary psychology approach, believe that natural selection did not and could not make all behavior adaptive. In this view, our brains could not anticipate *every* environmental challenge now and into the future, but evolved to respond to cues that were likely to enhance fitness in what has been called the "environment of evolutionary adaptiveness (EEA)." Thus, our brains were evolving a few hundred thousand years ago or more, when the human was a hunter-gatherer—not in the past ten thousand years, when we were inventing agriculture, religion, war, and other aspects of civilization. In other words, our responses to particular stimuli were selected out to enhance our fitness in a vastly different environment from today's, when drug pushers lurk on every street corner. Using this approach, I think we can grasp much greater understanding of both the biological force and limits on many current maladaptive behaviors.

For example, we like sweet food which, in EEA times, was found in ripe fruit and is good for humans. Then, a couple of hundred years ago, somebody figured out how to make candy, and—bingo!—a lot of people now prefer a Mars bar to grapes, a less fitness maximizing, and hence less adaptive, choice. Darwin would have predicted this since we are not selected to respond directly to fitness, for there is no way our genes could have prepared for candy or to adjust to its potential to rot teeth. We were selected to like only sweet things, and modern technology fools the system.

Or, take erotic pornography, which, though surely not fitness enhancing, does exploit the cues that in the EEA were triggers for reproductive success. Modern technology and culture pander to these systems that evolved in our brain to get another job done. Based on Darwinian principles, modern pleasure technology ought to be fitness neutral or fitness reducing. Perhaps in a few hundred thousand years candy and pornography won't excite people. Yet for primitive humans who evolved in the EEA, the fundamental stimuli concentrated in these events were fitness enhancing and, as a result, modern man is stuck with them. Our culture digs around to find pleasure buttons and pushes them, again and again.

There is another curious aspect to how most modern humans respond to these new pleasure technologies; most of us use them in moderation and do not suffer from addictions. When all is said and done, most humans don't walk around half crazed and victims to the pandering. Why is this the case?

The complex problem of addiction has been contemplated since the age of Aristotle, but only recently have investigators begun to take a scientific approach to it. In the past, the physical consequences of addiction to say, alcohol, have been hotly debated: Some characterize the addict as one who experiences problem drinking in conjunction with tolerance, withdrawal symptoms, and an inability to abstain; others have expounded upon this definition by stressing that an addiction is a compulsive behavioral pattern whereby one feels driven to seek a particular substance. Although there is no universally accepted definition of addiction, most definitions include components of both of these ideas, with the emphasis on compulsion. There seems little reason to doubt that owing to selection pressures all human brains react similarly to certain substances. But are the brains of those individuals who fall victim to addictions significantly different from the brains of those who do not?

There is now much evidence that there is indeed a brain-based correlate in people who become addicts. A growing body of research provides strong evidence for a hereditary link to many addictions. Some investigators see the biological factors that lead to a vulnerability as a "black box," which contains such variables as biochemical, genetic, and neurophysiological elements that may predispose an individual to addiction; but actual knowledge on the topic is sparse. At a fundamental level, it is not clear whether, for example, the genetic factors are directed toward specific addicting agents or, instead, control personality variables. In the latter case, for instance, one who inherits a gene particular for anxiety could be attracted toward a drug such as alcohol as a form of self-medication; while the alcohol might abate the symptoms of anxiety temporarily, in the long run, its use would likely develop into misuse, leading to "alcoholism." Since, however, the gene that started the whole cycle would have nothing to do with alcohol per se, it would be difficult to attach a particular behavior to a specific gene. This complication makes it very difficult to determine what is awry when one has a severe addiction.

The problem of addiction must be understood in connection with the history of drug use, the neurobiology of drug action on the brain, the prevalence of use, and any genetic disposition to abuse drugs. This framework allows us to bypass the histrionics that usually accompany discussions of the drug problem in America today.

Drug Use in the United States

Most of the drugs that today are illegal and considered highly dangerous and addictive were at one time commonly used for medicinal purposes or for pleasure in this country. Opium was in common use here in the nineteenth century. Originally, pure, cheap cocaine was widely distributed as a general tonic for sinusitis and hay fever and as a cure for morphine, opium, and alcohol habits; it was also an ingredient in medicines, sodas, wines, and cigarettes. The initial fear regarding cocaine originated in the South among whites who claimed that cocaine use caused blacks to rise up against the whites—an untruth that served as a tool of black repression. This fear of drugs and its effects on minorities has been a theme of antidrug sentiments throughout the history of this country (even though, when it was to their benefit, the middle and upper classes originally used cocaine to keep up the productivity of their workers). This checkered history of how Americans felt about drug use was further complicated by feeble attempts to control behavior in the first half of this century. There was no effort at federal regulation until after 1900.

Today, although Americans are frantic about the issue of drugs and addiction, they collectively consume huge amounts of drugs, both legally and illegally. At the height of drug consumption in the late 1970s, Americans consumed 8,000 tons of Valium, legally prescribed by the medical profession. The same drug, sold on the street, is labeled illegal and considered antisocial. Statistics indicate that, as recently as 1990, 6.9 billion gallons of alcohol were legally consumed in this country. Clearly, whatever we may say to the contrary, Americans like to alter their mental states. And Americans, of course, are no different from anyone else. But before discussing how this problem may be explained by selection theory, it must be explained exactly how drugs act on the brain.

Brain Chemistry and Psychological Rewards

Drugs act on the brain in ways that arouse or relax certain reactions, depending of the nature of the compound. Stimulants cause arousal and a sense of alertness, whereas depressants can produce relaxation and a sense of well-being. The brain is ready for these chemicals and receives, processes, and dispenses with their action in short order. All this brain activity is based on well-characterized cellular mechanisms. Every member of the human race has these mechanisms, and we have them because we use these same systems every day for the normal management of our psychological lives.

People take most psychoactive drugs for their rewarding psychological effect. If one feels tired and apathetic, stimulants work on the parts of the brain that cause activation; that mental boost is rewarding. If one is anxious, other drugs, like alcohol, work on still other parts of the brain that cause inhibition—an effect that is also perceived as rewarding. If one has none of these initial states but merely wants a psychologically pleasurable sensation, that too is easily achieved by psychoactive drugs.

Floyd Bloom and George Koob of the Scripps Medical Clinic in California have studied the effects of drugs on the brain for a long time. In the late 1980s, they proposed a mechanism of drug action for the three most psychoactive drugs (fig. 7.1) and identified brain networks that feed into a crucial brain nucleus, a nucleus known to innervate other brain areas in a way that gives rise to the sense of reward. Normally these brain systems are there to enable our personal sensations of reward as derived from our behavior in some social context. When one feels the normal course of events is not providing the right inner balance, one takes a small amount of some drug to supply the missing sensation and reward. That most of us do this in moderation is evident from the U.S. government's statistics on the prevalence of drug use (fig. 7.2).

Contrary to what many people assume, most Americans are not misusing drugs. Most Americans are exposed to drugs of all kinds during their lives but, in fact, refrain from indulging themselves in intense drug use. According to the National Household Survey (NHS), 90 million Americans have experimented with drugs at some point; in terms of availability, drugs might just as well be legal. But, although 90 million (or more) Americans have experimented with drugs, drug abuse is still at a (relatively) low level. Excluding tobacco, about 10 percent of the adult population in the United States abuses drugs. That figure, which includes alcohol, remains fairly constant. In our culture alone, between 70 and 80

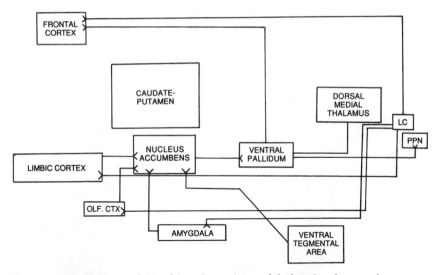

FIGURE 7.1 Bloom and Koob's schematic model showing how opiates, psychostimulants, and ethanol affect the brain's reward system. In this diagram, LC represents the locus ceruleus; PPN, the pendunculopontine nucleus; and OLF. CTX, the olfactory cortex. The nucleus accumbens, part of the basal ganglia, is the target for many of the dopaminergic links in the network. Reprinted from George F. Koob and Floyd E. Bloom, "Cellular and Molecular Mechanisms of Drug Dependence," *Science* 242 (1988): 721, figure 4. Copyright 1988 by the AAAS. Used by permission, with the consent of the author.

percent of adults use alcohol; the abuse rate is currently estimated at 5 to 6 percent. Clearly, the instinct to moderation is a major feature of the human response to drug availability. While most people intent on living productive lives enjoy the sensations of euphoria, anxiety reduction, and (at times) social disinhibition or even anesthesia, the desire for these sensations does not dominate behavior. Alcohol fills these needs for many people, and they generally manage it successfully. Early exposure to alcohol is common and inevitable. Yet studies have shown that it is difficult to determine which drunk at the college party will evolve into a serious alcoholic. What is known is that most early drinkers stop excessive drinking of their own accord without treatment.

The actual risk for individual drugs is much lower than the inclusive 10 percent figure. According to the NHS, an estimated twenty-one million Americans have tried cocaine, but only three million reported having used the drug at least once during the month preceding their interview. Most of the three million are casual users. All the cocaine users make up only 2 percent of the adult population, and the addicts make up less than

143

Prevalence of Drug Use in the United States in 1990 for the Age Group 18-25

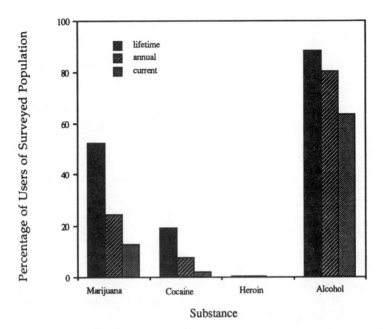

FIGURE 7.2   The prevalence of use of four major drugs in America during 1990 in the 18–25 age group. Adapted from The National Institute on Drug Abuse, issued by the Press Office of the National Institute on Drug Abuse: *Overview of the 1990 National Household Survey on Drug Abuse,* Table 1. Used with permission.

one quarter of 1 percent of the total population. Not exactly epidemic figures. Nor is the *rate* of addiction going up. There are fads where one drug becomes more popular than another; but when one goes up in consumption, others go down. Heroin use has gone down, and so has marijuana use. In fact, according to the NHS, use of cocaine and all other illicit drugs is on the decline among adults. The recent National High School Survey conducted by the University of Michigan reports the same is true among high school students.

Drug Addiction and Selection Theory

It is clear that the vertebrate brain comes equipped with mechanisms that help alleviate feelings like pain. Special receptors sprinkled throughout the structure respond to self-produced opiates called "endorphins." Since animals can also be anxious, depressed, tired, and blue, other receptor systems have come to exist, and chemicals in the body are produced to help modulate these moods. It is easy to imagine that selection pressures would establish these systems throughout all kinds of brains. Life is definitely better with them.

Since these effects are mediated by receptors in the brain, and since the number of receptors is determined by genetic mechanisms, it would follow that there should be variation in the number of these receptors and in the efficiency of their mechanism of action. Indeed, essentially all mood states reflect activity and variance in receptor mechanisms. Clearly, disorders of this system ought to have powerful consequences for behavior.

All of this suggests that a small percentage of us could possess certain subsystems in the brain that respond to exposure to particular stimuli so as to leave the individual with no easy way to refrain from overconsumption. This is the "there-is-a-genetically-driven-susceptibility-to-certain-kinds-of-substances" argument. On the other hand, certain brains may be driven by inherited personality traits, leading one to take drugs for relief. This is the "what-you-see-is-not-what-you-get" argument. Both arguments assume that there is a genetic component in the person using drugs, and, at the same time, a recognition by the person that they should not be doing so. This combination frequently produces powerful psychological reactions, such as guilt and loss of self-esteem, which may heighten one's sense of anxiety and thus feed back into the addiction, and so on. Selection theory would predict that the recidivism rate would be high.

Various psychosocial and cultural factors are thought to impinge upon the biological black box, either to trigger or to inhibit the biological vulnerability to addiction within. These factors include availability of the drug, acceptability of drug use (as within a culture or a peer group), the risk-taking propensity of the individual, and the social-reinforcing properties of the drug. The interaction between one's genetic and biological endowment and the environment can lead to varying outcomes. For example, a person with low or no biological vulnerability to addiction may nonetheless become addicted if the environment influences trigger the appropriate behavior. Thus, drug use increases with unemployment, as if one were driven to supply by artificial means the daily or weekly allotment of rewards we all seek. In this case, as soon as work is found, drug use stops.

On the other hand, a person with a high biological vulnerability to alcoholism will never become addicted if he or she never takes that first drink. It is more likely, of course, that the highly vulnerable person will become addicted to something or to many things. While some researchers, like Sanford Peele, have found the notion of a biological basis too reductionist for addiction, and emphasize social influences instead, the growing evidence from a wide range of research makes it virtually impossible to ignore the biological and genetic components associated with the hard-core addiction of the few Americans who are heavy users of drugs. But again, we do not understood whether the biological component is expressing itself directly by establishing in the brains of some people a predisposition for substances; or whether it is indirect and, thus, contributes to certain psychopathological traits that dispose one to develop dependency of one kind or another. These are crucial distinctions.

Alcoholism and Selection Theory

Alcohol is undoubtedly one of the most culturally accepted and widely consumed addictive drugs in the Western world. Ninety percent of Americans, for example, drink at some time. Two thirds of all men and slightly fewer women will have at least one drink during any given year. Ninety-two percent of high school seniors had tried alcohol in 1988, and nearly two thirds were current users. The statistics for alcohol abuse, however, show a rate of abuse of about 6 percent of those who drink. This hard-core group represents a lot of people, however, and it is they whom we are trying to understand in terms of selection theory.

Alcohol research in recent years has suggested that there may be subclassifications of alcoholism, differentiated by causal factors and behavioral patterns. Robert Cloninger, a psychiatrist from Washington University in St. Louis, believes that the extent to which genetics and environment are contributing factors in the etiology of the disease depends on the subtype of alcoholism. He suggests that there are two types of alcoholism. Type-1 alcoholism, also known as "milieu-limited" alcoholism, involves both a genetic predisposition and environmental provocation toward alcoholism. Type-1 alcoholism occurs in both women and men, and the alcoholic is characterized by anxious personality traits and dependence on the anti-anxiety effects of alcohol. In contrast, Type-2 alcoholism is determined almost exclusively by genetic factors. It occurs only in men, and the Type-2 alcoholic is characterized by antisocial personality traits, criminal activity, and the pursuit of alcohol for its euphoric effects. Because of the high heritability of Type-2 alcoholism, environmental factors are not supposed to prevent the manifestation of the disease: Rather, the environment can only modify the severity of the condition.

The evidence for a genetic link in alcoholism, particularly in Type 2, has been accumulating. Much of the research has been done through studies of families with a history of alcoholism. In an analysis of thirty-nine studies of familial alcoholism, it was found that 25 percent of alcoholics had alcoholic fathers. Similarly, family and adoption studies have suggested that 20 to 25 percent of sons of alcoholics become alcoholics. In general, the offspring of an alcoholic parent are two to four times more likely to become an alcoholic than are children of nonalcoholic parents. However, family studies cannot be considered conclusive evidence for a strictly genetic transmission of alcoholism since families share not only common genes but a common environment as well. In order to zero in on the genetic component per se, twins studies and studies on adopted children are necessary. Adoption studies are useful in that they separate the genetic component of a family from the environmental component. These studies found that the incidence of alcoholism correlated with the biological parents rather than with the nonbiological rearing parents. Donald Goodwin, from the National Institute of Health (NIH), after reviewing numerous Danish adoption studies, concludes that alcoholism is four times more likely to occur when the biological parents are alcoholics than when the nonbiological parents are alcoholics. Even brothers reared apart had the same incidence of alcoholism.

Twin studies have also proven to be a useful tool in the investigation of familial transmission of alcoholism. Monozygotic twins share all of the same genes, as opposed to dizygotic twins, who share only 50 percent of their genes (the equivalent of any sibling). Several studies have found that monozygotic twins have a higher concordance rate for alcoholism than do dizygotic twins. Since environmental factors presumably are no more variable for dizygotic twins than for monozygotic twins, the studies indicate that the genetic similarity between monozygotic twins is responsible for the high concordance rate in alcoholism, thus emphasizing its alcoholism.

Recently a new wave of research has focused on specific genetic markers that indicate a susceptibility to alcoholism. Sons of alcoholics, although not alcoholic themselves, have been shown to exhibit behavioral and physiological differences in response to alcohol in comparison with sons of nonalcoholics. For example, it was found that children of alcoholics reported lesser feelings of intoxication and exhibited less body sway in response to alcohol than did children of nonalcoholics. Presumably, the bodies of people who have some genetic predisposition for alcoholism somehow respond differently to alcohol.

Electrophysiological markers have also shown differences between sons of alcoholics and sons of nonalcoholics. EEG testing, or brain-wave analysis, has shown that alcoholics have fewer slow waves (alpha waves) before consuming alcohol, but show a greater increase in slow waves in response to alcohol consumption. This condition was once thought to be a result of years of alcohol abuse, until it was demonstrated that sons of alcoholics also tend to show the same increase in alpha waves after intoxication. Alcoholics and sons of alcoholics also show a difference on another brain-wave measure, the P300. Here, alcoholics and their sons have a lower P300 wave than do normals. Again, the idea suggested in all of these studies is that the alcoholic's brain is somehow different.

Once again, however, caution is needed in evaluating these studies. The brain-wave abnormality seen in alcoholics is also seen in patients with various mental disorders. Although studies like these point to a biological dimension to an addiction, they have not yet unearthed a biological correlate of alcoholism per se.

Biochemical markers for alcoholism have also been the subject of extensive research. Lower MAO platelet levels have been found in Type-2 alcoholics (the more genetically transmitted subtype of alcoholism) in comparison with Type-1 alcoholics and normals. Currently a debate is raging over whether alcoholism might be due, in part, to the activity of something called the D_2 dopamine receptors (fig. 7.3). It is

FIGURE 7.3   A schematic view of the possible sites in the brain where dopamine works to modulate various drug actions. Adapted from J. R. Cooper, Floyd E. Bloom, and Robert H. Roth, *The Biochemical Basis of Neuropharmacology*, 4th ed. (Oxford: Oxford University Press, 1982), with permission.

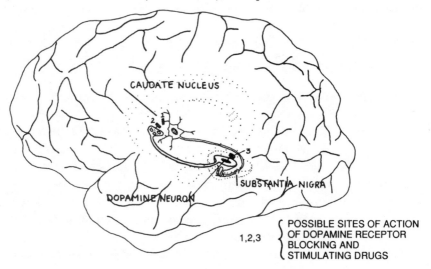

known that when drugs that block the action of this receptor are given to animals, the animals consume more ethanol. It has therefore been suggested that regulation of this receptor is involved in alcoholism. It has recently been argued that severe—and I mean truly severe—alcoholics may have disorders in the gene responsible for the D_2 dopamine receptor. If this proves to be true, it would suggest that variation in genetic expression gives rise to an unabated and immoderate use of a substance normally regulated by our species brain. This converging evidence suggests a genetic basis for at least some subtypes of alcoholism.

In all of this, however, it is important to keep emphasizing the caveats. Rather than talking about the "genetics" of alcoholism or the "biochemistry" of alcoholism per se, one should really talk about the genetics of factors that lead to the use of alcohol. This properly leaves open what it is that is being selected out in someone's biological system when mood-altering substances are being abused.

The genetic story has another side as well. Not only may there be a genetic susceptibility to alcoholism, but it has also been argued that there is evidence for genetic *protection* against alcoholism. Asians have long been known to have a low incidence of alcoholism. It is no coincidence that 30 to 50 percent of Asians are missing the ALDH isoenzyme, which leads to higher blood acetaldehyde when drinking. This causes the "oriental flush" (redness in the face) and unpleasant side effects such as palpitations and nausea. The question becomes whether the low alcoholism rate for Asians is a result of genetic or cultural factors, and its answer provides a warning against simplistic interpretation of any drug data.

The Chinese serve as a good example of how biological and cultural variables play out. Chinese babies who are given a drop of port with their milk will flush red, whereas occidentals will not. Adult Chinese, after drinking beer, report tachycardia, dizziness, sleepiness, and a "hot stomach." The fact that a dose of alcohol leaves an occidental unaffected but overwhelms a Chinese subject with a physiological response seems to provide an easy rationale for why alcoholism is a minor problem among Asians. As Professor Stanley Schachter of Columbia University has pointed out, the differences could be attributed to the fact that Asians may go to sleep before they become violent or offensive, or go to sleep too soon to drink too much alcohol, and so on.

But Professor Schachter cleverly reminds us of the Asiatic groups who made their way to America—the North American Indian and Eskimo. Both of these groups are notorious for having drinking problems. Why is this so? Although studies have shown that the American Indian and the Eskimo respond physiologically the same way Asians do, we do not yet fully understand why they have such problems with alcohol abuse. However, it is clear that social and cultural factors can override a predisposition, or lack thereof, to drug abuse—an important point to keep in mind, especially when considering other data concerning how alcohol may trigger certain reflexes in our physiology. Consider the following.

Selective breeding has enabled scientists to develop a strain of rats that actually prefer an alcohol solution when given a choice between alcohol or water. These animals will voluntarily ingest alcohol to the point of intoxication and physical dependence. Alcohol-preferring (P) rats develop a tolerance more quickly and a longer lasting one than do non-alcohol-preferring (NP) rats. P rats may acquire tolerance to alcohol for up to ten days from a single dose. A key physiological difference between the breeds can be found in the neurotransmitters (that is, norepinephrine, dopamine, serotonin, and GABA). In particular, P rats have less serotonin and dopamine than NP rats.

Strains of mice with different behavioral effects from alcohol have also been developed. Short-sleep (SS) mice are those that show less severe effects from alcohol than do long-sleep (LS) mice. The classic difference is that SS mice return to their alert "righting" position more quickly than do LS mice. Again, differences in neurotransmitters, particularly serotonin and GABA, may be linked to alcohol sensitivity. Thus, it appears that a predisposition or vulnerability toward alcoholism is transmitted from one generation to the next.

I have focused on alcohol because much is known about the abuse of this drug. Other compulsive drug-seeking could well be related to other biological factors. As mentioned before, many drugs of abuse act on the reward system of the brain. George Koob has recently reported that each of these drugs may affect the reward system through different channels. Thus, one could block the action of a particular drug that has a reinforcing effect, and yet not shut down the whole reinforcement system of the brain.

Everyone in our free society has access to drugs and alcohol. Some people would like to have all these substances removed from our lives so as to prevent the small percentage of us who could fall victim to them from running that risk. Yet it is far more reasonable to live with the dominant fact of drug use. Our species has mechanisms that allow us both to enjoy drugs and to regulate their moderate use. Parents who find out that their child has been exposed to a drug should take comfort in the overwhelming odds that it will not dictate the child's life. If addiction should result, they ought to look for other associated problems and begin intense education on the necessity for abstinence.

COMPULSION AND OTHER EXTREME BEHAVIORS

When addictions are seen as possibly providing a sort of time-out from an otherwise uncomfortable mental state, the door is open to consider addictive behavior a general strategy for relief, and assume that people pick different remedies or diversions. In this light, perhaps other exaggerated behaviors can be better understood as also having some kind of underlying biological substrate.

Although the term *addiction* is most often used in reference to a physiological and/or psychological dependence on a chemical substance such as alcohol or nicotine, it is not uncommonly also used to refer to a particular behavior, such as hypersexual activity or gambling. Thus, groups such as Gamblers Anonymous and Sex & Love Addicts Anonymous see a particular vice as a disease, claiming that they are unable to engage in the behavior in moderation or with restraint. Is it possible to become addicted to a behavior as one does to a drug? Does some predetermined brain circuit promote and encourage excessive behavior in some individuals?

In trying to understand such behaviors in the context of selection theory, I am moving into dangerous territory. Many an evolutionary biologist would, though believing strongly in nature, nonetheless dismiss extreme human behavior as simply an aberration induced by culture. This sentiment is certainly well founded and one for which there is abundant support. Human males and females, for example, seem to have reversed the general biological rule about advertisement. In the animal kingdom, it is the males, not the females, who adorn themselves and call out for sexual partners. In our culture, it is usually the males who are drab and unadorned while females use makeup, false eyelashes, and other devices to make themselves more colorful and alluring.

Sexual activity pushes one of our pleasure buttons. The selection pressures for enjoying sex have been enormous. What happens in extreme cases? When sexual activity becomes extreme, does it reflect a cultural influence or an overactive biological system—an accidental variant on the one that has been selected for? Here is the testimony of a so-called sex addict:

Q. When and why did you feel that your sexual behavior was a problem?
A. I knew all along that my sexuality was a problem—back as far as high school. I didn't realize that I was a sex addict until I stopped drinking and doing drugs. I was in Alcoholics Anonymous (AA) at the time. I realized that I had to stop having sex or I would start drinking again. I was using

sex with men to avoid dealing with my sexual feelings about women. I decided to go to Sexual Compulsives Anonymous (SCA).

Q. How did you hear about SCA?

A. From the guys at AA. They talked about it. I knew about it for a year before I finally decided to go.

Q. Have you any other addictions?

A. Yes, I've spent my whole life juggling my addictions to stay alive. I went to Overeaters Anonymous (OA) first for bulimia. I was drinking at the time. I thought that I wasn't an alcoholic if didn't drink alone because my mother, who was an alcoholic, always drank alone. Then I was sent to AA by OA. For years I substituted one addiction for another. I've been addicted to alcohol, drugs, sex, food, caffeine, cigarettes, shopping, and gambling.

(SIECUS Reports, pp. 1–3)

The currently accepted definition of addiction with its emphasis on compulsive behavior makes it seem that hypersexuality and gambling are also addictions in some sense; and certainly this person saw all of the behaviors as part of a whole pattern. Addictions, thus, seem to draw upon another variable of our personalities—namely, compulsions. An obsessive-compulsive disorder, according to DSM-III, is characterized by "repetitive and seemingly purposeful behaviors that are performed according to certain rules or in a stereotyped fashion. The obsessions or compulsions are a significant source of distress to the individual or interfere with social or role functioning." Although most often applied to ritualistic thoughts and behaviors such as cleaning or count-ing, perhaps obsessive-compulsive disorder (OCD) can include exces-sive involvement in otherwise socially accepted behaviors (like sex and gambling). By placing these seemingly learned behaviors in the context of OCD, we can continue to search for possible organic causes of the phenomenon.

As with most emotional disorders, a primary issue is the etiology of the obsessive-compulsive disorder. Does it emerge as a result of a flawed upbringing or inappropriate environmental reinforcement? Or are biolog-ical elements at the root of the behavior? Additionally, if sex and gam-bling are to be considered examples of obsessive-compulsive behavior, can similar biological substrates be found for excessive involvement?

Let's first consider the evidence for a biological substrate of OCD. Once considered a rare psychological disorder, OCD is now estimated to occur in as much as 2 to 3 percent of the general population; with a subcortical structure commonly associated with motor function, the basal ganglia, most often implicated (fig. 7.4). Numerous studies have distin-

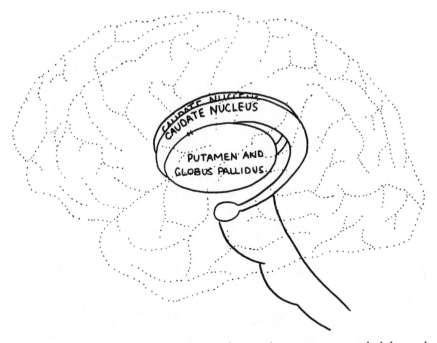

FIGURE 7.4 The basal ganglia. The caudate nucleus, putamen, and globus pallidus, together with portions of the thalamus and other nuclei, work with the cerebral cortex to control both motor and sensory functions. Adapted from Arthur C. Guyton, *Basic Neuroscience: Anatomy and Physiology* (Philadelphia: W. B. Saunders, 1987), fig. 2.7, p. 15. Used by permission.

guished physiological differences in this circuit when comparing OCD patients with normal controls. It was revealed by CT scans that patients with OCD have a smaller volume of some of the brain nuclei within the basal ganglia than normals. PET-scan studies have demonstrated an increased metabolic rate for this region and, in the frontal lobe, and cingulate pathway of OCD patients. These results suggest a relationship between OCD and Georges Gilles de la Tourette syndrome, a disorder of recurrent motor and phonic tics that is also linked to basal ganglia abnormalities. The fact that approximately 20 percent of OCD patients have facial tics and between 12 to 35 percent of patients with Tourette's syndrome also have OCD, reinforces the relationship between these two disorders and also provides additional evidence for the role of the basal ganglia in OCD. It has been suggested that OCD may represent an alternative expression of factors responsible for Tourette's syndrome, since they appear to share common physiological substrates. Evidence for this theory comes not only from patients with Tourette's syndrome and

OCD, but from their families, who show an increased incidence of both disorders.

Other investigations of familial history also suggest a biological or genetic factor in OCD. For instance, in one study of OCD patients, 5 percent of the parents had OCD, and 11 percent had significant obsessive-compulsive traits. Additionally, a higher concordance rate for OCD was found among monozygotic twins than among dizygotic twins.

Another reason that OCD patients are considered to have a biological etiology is that pharmacological treatments have proven to be effective. Thus far, the most successful treatments have been clomipramine, fluoxamine, and fluoxetine, which all block reuptake at the neural synapse of serotonin, a neurotransmitter active in the basal ganglia.

Thus, it is becoming increasingly more clear that obsessive-compulsive disorders do have biological roots and pharmacological solutions. But do these findings have any bearing on compulsive behaviors such as sex and gambling? Some investigators have suggested that compulsive features may be evidenced in a variety of behaviors such as sex, gambling, alcohol consumption, smoking, eating, and drug taking. Others maintain, however, that what society views as compulsive or abnormal behavior is merely a judgment passed on those who exceed the limits of the norm.

Hypersexuality

There are many terms to describe the "infliction" suffered by those who indulge in and exhibit excessive or inappropriate sexual behavior—nymphomania, Don Juanism, hypersexuality, to name just a few. These people engage in behaviors that are not considered normal in our culture, such as excessive promiscuity with inappropriate or arbitrary partners. Some researchers argue that these "hypersexuals" have merely been labeled by the values of our culture, time period, and therapists. For instance, homosexuality and nonnormative sexual behavior (such as masturbation and fellatio) were considered to be sexual disorders in the 1952 edition of the *Diagnostic and Statistical Manual of Mental Disorders* (DSM-I), whose authors asserted that society passes a judgment on those who engage in any activities that fall outside a normative standard.

The concept of sex as an addiction originated from members of Alcoholics Anonymous who applied AA's 12 Steps to Recovery to sexual behavior, forming Sex and Love Addicts Anonymous. Three levels of progressively more serious sexual addiction were defined: promiscuity, pornography, and so on; exhibitionism and voyeurism; and rape, incest, and child molestation. Those who consider sex an addiction assert that loneliness, low self-esteem, and anxiety cause individuals to lose control over their sexual behavior. Sex gives these individuals a temporary psychic relief—a sexual "fix."

Critics of a sexual addiction model cite numerous reasons that sex cannot be considered an addiction. First, sex is not a substance; it is an *experience,* and one does not become addicted to an experience. Second, there is no physiological state of dependence: that is, no withdrawal symptoms, such as diarrhea or delirium. Finally, if one were truly addicted, then one could never learn to have sex in moderation, yet hypersexuals are currently counseled to do so. If we define hypersexuality as lack of control over an erotic impulse, then we may conclude that "sexual compulsion" is the more appropriate term. Compulsive sexual behavior may be triggered by anxiety and loneliness, but the temporary relief from sex causes guilt, leading to increased anxiety, thus creating a vicious cycle. Interestingly, most perpetrators of rape and incest exhibit features of compulsion and not of paraphilic eroticism (deviant sex behavior). It is also evident that there are sex differences in the expression of sexual compulsion. Men tend to exhibit uncontrollable promiscuity, autoeroticism, voyeurism, and exhibitionism, and commit sexual offenses such as rape, molestation, and incest. Women engage in frequent dangerous sexual encounters.

While some psychologists argue that hypersexuality is not a medical condition but merely a stigma attached to nonnormative life styles, others suggest that a real biological substrate may lie at the root of hypersexual behavior. Jerome Goodman of Columbia University suggests that very sexually active people may simply be at the extreme end of the biological bell-shaped curve, and accepts the fact that hypersexuals do not suffer from a disorder per se, but may merely represent the fringes of the curve. In contrast, others believe that a neurochemical failure causes people to fall "heatedly in love, often with inappropriate partners." Evidence for a biological disposition toward excessive sex or love patterns is found in a correlation between hypersexuality and physical stature. Goodman found that hypersexuality in women is correlated with small breasts, a low voice, acne, and hirsutism. Further, hypersexuality is associated with psychiatric disorders known to have biological etiologies, such as schizophrenia and mania. Epilepsy, frontal lobe injury, and limbic and subcortical dysfunction have also been linked with hypersexual activity.

Gambling Behavior

The issue of life-style versus lack of personal control versus disease has also been explored in gambling research. In the recent DSM-III, pathological gambling is listed under the category "Disorders of impulse control not otherwise classified." It is characterized by "a chronic and progressive inability to resist the urge to gamble, which compromises, disrupts, or damages family, personal, and vocational pursuits and is not due to antecedent anti-social personality disorder. It is of unknown origin." Gambling is often considered an addictive behavior, when one cannot resist the impulse to gamble and does so in spite of the inevitable cost in terms of finances or family life, and comparisons are made to other models of addiction. Yet the fact remains that gambling is a behavioral pattern and not a substance, and therefore considered to be a disorder of impulse control rather than an addiction.

Nonetheless, research on gambling has provided evidence that biological factors may be involved in pathological gambling. Electroencephalogram (EEG) studies revealed that, in comparison with normal controls, pathological gamblers who had been abstinent for at least two years had a deficit in the degree of EEG differentiation produced by simple verbal versus nonverbal tasks. The EEG patterns displayed by the pathological gamblers were the same as those found in children with attention deficit disorder (ADD). A post hoc survey revealed that a significant number of compulsive gamblers showed symptoms of ADD as a child. A similar link has also been found between ADD and alcoholism. The relevance of ADD to compulsive and/or addictive disorders becomes even more enticing when one considers the primary characteristics of ADD: an inability to sustain attention, and impulsivity—the tendency to respond impetuously, without forethought, the relative inability to delay gratification, the tendency to respond even if the consequence may be negative. This suggests that disorders of impulse such as gambling may share a similar etiology with alcoholism and other addictions.

The serotonergic system once again appears to be a determining factor in pathological gambling (and in ADD as well). Deficits in serotonin have been shown to reduce attentional processes in both humans and laboratory animals. There is also evidence for a functional disturbance of the noradrenergic system in pathological gamblers, as evidenced by high urinary output of norepinephrine, high levels of centrally produced CSF MHPG, and low levels of plasma MHPG. The noradrenergic system may be responsible, in part, for regulating sensation-seeking behaviors.

Another interesting analogy between alcoholism and pathological gambling was made by Richard McCormick of the Cleveland Veterans Hospital, who identified numerous subtypes of pathological gamblers. Subtype A is the recurringly depressed pathological gambler who has a depressogenic cognitive imbalance and possibly a biochemical imbalance, but is not depressed simply as a result of his or her gambling problem. McCormick proposes that this type of gambler needs to gamble to relieve his depressive symptoms, but gambling increases his or her depression and anxiety, and thus a cycle of behavior emerges. The Subtype-A gambler is similar to the Type-1 alcoholic, who drinks to reduce anxiety and avoid life's problems. McCormick also describes the Subtype-B gambler, the chronically understimulated gambler. Like the Type-2 alcoholic, the Subtype-B gambler needs excitement and stimulation, has poor impulse control, and exhibits hyperactivity and low frustration tolerance. These subtypes may be important for future research in determining different etiologies of pathological gambling.

In reviewing the evidence, it appears that there are certain similarities between the DSM-III Obsessive-Compulsive Disorder, Pathological Gambling, Hypersexuality, and some addictions, such as Alcoholism. The similarities, for the most part, have to do with an inability to resist impulses. Evidence for a common biological etiology comes most strongly from abnormalities in EEG patterns and neurochemical regulation (especially serotonergic and noradrenergic systems). The notion of a generic addictive personality has not been borne out in the literature, and many personality traits may simply reflect the psychological state of a person trapped by particular behavioral patterns.

The final issue is whether these compulsive behaviors can be considered actual diseases. Some critics assert that we too easily label as a disease any condition that is difficult to regulate or control; and maintain that by calling alcoholism, pathological gambling, or hypersexuality a disease, we are freeing people from personal responsibility for their actions. Gamblers will be able to lose their money without guilt, and rapists and molesters will be able to commit crimes without blame. Clearly, the advocates of a disease model of addiction do not free people with addictions or compulsive behavior from personal responsibility to *treat* their illness. The disease model is used to emphasize that biological elements may play a role in the person's behavior, just as biology has been found to be a factor in schizophrenia, mania, and other mental disorders. The patient is not blamed for the disease, but is expected to treat and control his or her behaviors.

Others have argued that society uses such terms as *disease* or *abnormality* as a means of condemning a particular life-style. Though this argument merits consideration, it is clearly not valid for a large percentage of people with OCD. Most people who are diagnosed as hypersexual or as compulsive gamblers are not simply people with a nonnormative life-style. These are people who have not been able to control their behavior to the point of disrupting their personal and professional lives, and possibly causing harm to others. Their problem is that they want to change their life-styles and are unable to do so. This inability clearly indicates a disorder and not a value judgment of one's personal decisions. Finally, the disease model, in its purest form, rejects the notion that the afflicted can ever consume a substance or engage in an activity in moderation—a claim that accumulating research has, in fact, refuted. Recovered alcoholics have been taught to drink in moderation and hypersexuals have recovered from their condition.

Maladaptive behaviors are a bizarre manifestation of biological systems that have been selected out to promote survival of the species. The idea that emerges from the foregoing analysis is that brain systems selected for in the EEA can function abnormally and create internal body states that humans then attempt to adjust for. The adjustments, whether through drugs or dangerous behavior patterns, often create more problems than they solve.

CHAPTER 8

Selection Theory and the Death of Psychoanalysis

Ever since the early seventeenth century when Francis Bacon urged his colleagues to stop debating how many teeth a horse has and to go out and count them, science has taken its own course, a course that has brought true insight into the nature of nature. Each introduction of a method to measure a secret of a living process further demystifies it. Nonetheless, these advances always occur in the context of a prior theory about a process, and the purveyors of the theory not only hold desperately to their position but, not surprisingly, also attempt to fit the advances into their own positions. This tendency appears to be a property of our species; and no where is it more vigorous than in the field of psychoanalysis.

When it comes to the study of the human mind, theoretical positions are both well worked and well known. From a philosophical perspective, the arguments have raged for centuries. From the psychiatric perspective, this century has seen the rise, and recent fall, of psychoanalytic theory. Over the past thirty years, I have been fascinated with the boundless energy of the practioners of this art. They cannot seem to let go of their theoretical positions, and none of them is the least bit interested in getting out of the armchair and into the lab.

About a hundred years ago, when Freud was formulating his theories of mind, most of his contemporaries distinguished between biological factors and psychological processes. Yet although many scientists and clinicians of the time freely attributed certain mental disorders to biological phenomena, they had no real knowledge about brain genetics, brain chemistry, cellular organization of the brain, developmental abnormalities, or a host of other biological issues that are now commonplace. Freud himself said that "the deficiencies in our (the psychoanalysts') description would probably vanish if we were already in a position to replace the psychological terms by physiological or chemical ones. . . . Biology is truly a land of unlimited possibilities."

With the biological model then little more than a belief, it is no wonder that Freud spent his time considering possible psychological variables. In time, however, some of Freud's disciples took these variables to extremes. Some tried to explain all symptoms in terms of the "unconscious": Thus, "A sore throat was developed to force the whispering of secrets, pain in the arm to ward off a tendency to forcefulness or thievery . . . visual difficulties always expressed emotional conflicts and retinal bleeding and other organic changes in the eye were efforts to defend the patient against forbidden wishes." Everybody can and does attribute cause to antecedent psychological events. Elaborate on that general idea a little, and you can have a theory of mind.

Dismantling the Cathedral of Psychoanalysis

Today, of course, there are many theories of mind. Whereas neuropsychologists, as well as social and personality psychologists, find a cognitive approach to questions of mind most appealing, psychoanalysts cannot seem to agree on either a standard conception of the structure of mind or its inherent or characteristic processes. Psychologists see the mind as composed of cognitive maps, and of perceptual, linguistic, and memory processes. Psychoanalysts, on the other hand, are compelled to look upon the mind as a collection of ideas, impulses, and desires, and, more literally, as "motivating representations of people and parts of people—particularly breasts and penises." The object of study in each case seems hardly to be one and the same animal.

In his landmark book, *The Selfish Gene,* Richard Dawkins points out that in addition to biological evolution, which is driven by the blind purpose of gene survival, there is cultural evolution. He dubs his unit of analysis the "meme." Memes are ideas, customs, and traits that replicate and are, like genes, passed from one generation to another. Memes are also constrained and must compete for survival, just like genes. Freudian ideas can be considered good memes because they replicate with abandon and have clearly survived.

The psychoanalytical theories of Freud and others have persisted for a variety of reasons, primarily because they offer a system that attempts to explain the inexplicable! Husband kills wife, or vice versa. Or, worse, mother kills son, or vice versa. At a less drastic level, moods like depression, anxiety, or compulsiveness are seen as being driven by unconscious and suppressed desires, wishes, and so on. We humans are fascinated with these phenomena and want an explanation. When such events actually occur, it takes us about two minutes to cook up an explanation that might account for the event. Inevitably, those explanations are psychological in nature.

Once in place, of course, personal psychodynamic thinking was further validated in modern culture. Freud's ideas are marketed by untold thousands of practitioners. They are also reinforced when used in their "popular" form. If you notice someone being defensive about something he or she has said or done, you are likely to call such behavior "Freudian"— even though that same behavior may be explained by any one of a dozen ideas about the psychological nature of humans.

Another reason Freudian ideas have gained such hold is that people who are drawn to psychoanalysis and Freudian ways of thinking about psychological process seem to have a predisposition to become fascinated with a formal analysis of psychological states of mind: That is, the client who selects psychoanalytic analysis is very likely to believe in it. As a rule, such people are highly verbal and enjoy verbal constructs for just about everything. The verbally inclined tend to believe that all matters can be articulated, including states of mind. Drawing distinctions, or creating an explanatory framework that, on the surface, appears to explain a psychological event is all that is needed. Even so, simply asking anyone undergoing psychoanalysis to explain his or her behavior, does not produce a great volume of information for the analyst. The person will not deliberately lie, but will frequently make mistakes when asked to guess at causes.

For all of the shortcomings of his theory, Freud was indisputably an intellectual giant, a man of great insight and wisdom. Yet, in spite of his insight, Freud failed to consider adequately mental processes in biological terms, and view them in evolutionary terms. In recent years, two evolution theorists, Martin Daly and Margo Wilson, have convincingly laid to rest some of Freud's most cherished constructs, including the most renowned, the Oedipal conflict. Their account is as follows.

The Oedipal Conflict

The Oedipal conflict, known to most of the modern Western world in some form or another, takes its name from the events of the ancient Greek play *Oedipus Rex*. In the play, it is revealed by an oracle that Oedipus, though at that time still unborn, will someday murder his father, King Laius, and marry his mother, Jocasta. To prevent the fulfillment of the prophecy, Oedipus is banished from the kingdom at birth and is raised as a prince in an alien court. As he grows older, Oedipus doubts his origin, and, in seeking the truth about himself, discovers the awful events foretold by the oracle. Attempting to escape his destiny, Oedipus flees from his adopted homeland and the people he believes to be his mother and father. In doing so, Oedipus heads directly toward his true home, and in a series of events that follow, unwittingly murders his father and marries his mother, thus fulfilling the oracle. Years later, when the truth of these events unfolds, Oedipus blinds himself and casts himself out of his homeland.

Freud's interpretation of this tragedy gave rise to his Oedipal theory. In this theory, Freud translates the play's events into the language of psychosexual development. Freud believed that Oedipus's murder of his father and subsequent marriage to his mother represented the fulfillment of repressed childhood fantasies—fantasies in which the father is viewed by the child as a sexual competitor for the mother's attentions. As Freud saw it, audiences were able to fulfill their hidden needs vicariously by watching Oedipus act out their collective, primeval wish. Beyond this, audiences could also be sure that Oedipus would receive punishment for his atrocity, and thus conform to society's rules concerning parent-child relationships. Individuals who succeeded in balancing the competing drives of wish fulfillment and societal constraints were considered to be managing their Oedipal conflict. Those who did not became psychoneurotic and required the help of a trained analyst.

To Freud, the play marked not only a description of events that are played out in the subconscious, but also the fact that the true significance of these events are, by and large, hidden from the individual. Had Oedipus known who his parents really were, he would have avoided fulfilling the prophecy, and thus avoided untold suffering. It was this hidden aspect of the play's events, and, by extension, the hidden aspects of the subconscious, that defined the analyst's role—the enlightened analyst could point out the "Truth" about mankind's hidden urges and eventually lead humanity to the path of true happiness. As Freud points out in his classic *The Interpretation of Dreams,* "The action of the play consists in nothing other than the process of revealing, with cunning delays and ever-mounting excitement—a process that can be likened to the work of a psychoanalysis." This perspective makes for wonderful literary theory, but is entirely out of step with biological science and selection theory.

One would think such ideas would die a happy death as modern science accumulates more sensible ideas about how the brain subserves behavior. Yet this is not the case. As recently as 1982, Melford Spiro wrote a book about how a variant of the Oedipal complex must be played out among the Trobriand Islanders. In a Trobriand boy's life, the significant adult male is not his father, but his mother's brother. In spite of this, Spiro sees the manifestations of Oedipal conflict everywhere. The Oedipal theory "predicts" that men would be very jealous of the women they are attached to, and simultaneously wish to possess women who are already attached to other men. Spiro argues that the son's frustrated wish for an exclusive relationship with his mother fuels the fire of a repressed Oedipus complex, which sets the tone for all future relationships. Needless to say, this prediction of male sexual rivalry is upheld. But, as Daly and Wilson point out, the same prediction is just as well fulfilled in other mammalian males—from a bull seal to a billy goat. Males are sexual competitors for reasons that have nothing to do with the relationship between immature son and mature father.

But while old ideas die hard in the scientific community, they are even harder to kill in the popular psyche. Members of our society still commonly assume that kin kill each other, citing the often-quoted line about how it is safer to be in Central Park alone at 3 A.M. than in one's own bedroom. The same belief is expressed in the work of Richard Gelles and Murray Straus, the best-known investigators of family violence:

> The family is the most frequent single locus of all types of violence ranging from slaps, to beatings, to torture, to murder. Students of homicide are well aware that more murder victims are members of the same family than any other category of murder-victim relationship. . . . In fact, violence is so common in the family that we have said it is as least as typical of family relations as is love.
>
> (Gelles and Strauss, 1979, p. 188)

If this is true, homicide is a human activity that defies the principles of natural selection. Daly and Wilson asked whether there was actually anything in the data to substantiate such a position. Their results offer a compelling and convincing conclusion that blood relatives generally do not kill each other. They start with the Detroit police statistics and then tackle a host of related claims about kin killing, such as infanticide, suicide, and spite killings. The Detroit statistics, as well as other statistics collected in Canada in the early 1970s, show that of the 512 homicides committed in 1972 that had been resolved by 1980 (fig. 8.1), 502 had been solved, and the killer was known. Of those cases, 243 were carried out by unrelated acquaintances (47.8%), 138 by strangers (27.2%), and 127 (25%) by relatives. Of the relatives, only 32 were consanguineal, the rest being husband and wife in-laws and so on. Thus, of the overall sample, only 6.3 percent were true kin—a figure hardly in line with the Oedipal theory, but consistent with selection theory.

After a thorough analysis of all kinds of behavior—homicide, infanticide, suicide, and so on—in our own as well as other cultures, Daly and Wilson reveal a simple truth: Blood killings are extremely rare and, when they do occur, they have explanations in keeping with selection theory.

FIGURE 8.1 A study related to that conducted in Detroit shows the homicide risk to children from a natural parent and from a nonrelative in Canada (1974–83). Contrary to public perception, most homicides do not involve kin. Adapted from Martin Daly and Margo Wilson, *Homicide* (New York: Aldine de Gruyter, 1988), pp. 76–77.

Risk of Homicide by Natural Parent or Nonrelative in Relation to Child's Age in Canada (1974-1983)

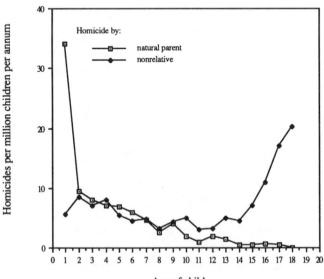

Thus, it seems that once Oedipal theory was suggested, and, its "existence" was established in psychiatry, it took on its own reality and became part of our thinking about the nature of mind.

This kind of phenomenon is not unique, as has been demonstrated by Henri Zukier, a social psychologist at the New School for Social Research. He was fascinated with the problem of the "Holocaust syndrome," in which grandchildren of victims of the Holocaust were deeply affected by the horror of those events. Zukier began to examine the issue and traveled to Europe and elsewhere to find similar cases and to discuss the problem with other psychiatrists. To Zukier's astonishment, the syndrome did not exist in Europe. He found the "disease" to be a construct of American, and in particular of New York, psychiatry. Perceiving an everyday neurosis in the context of a patient's family history, they created a label and began to apply it to other descendants of Holocaust victims. Soon, it had its own reality, a reality that bore no resemblance to anything other than a psychiatrist's imagination.

Tourette's Syndrome

This process is rampant in the field of psychiatry. Consider Georges Gilles de la Tourette syndrome. New Yorkers are all too familiar with the poor souls who mutter obscenities to themselves and shake from every joint with tics; who grunt, bark, and commonly involuntarily repeat the last words of others. In 1921, Sandor Ferenczi declared this a puzzle for analysts and for thirty-five years, they slugged away at it with such characterizations as "stereotyped equivalents of onanism and . . . the remarkable connection of tics with coprolalia . . . might be nothing else than the uttered expression of the same erotic emotion usually abreacted in symbolic movements."

Not to be outdone, another psychiatrist disagreed and instead offered: "The tic which takes the form of making grimaces has an obvious hostile significance. . . . Other tics, particularly coprolalia show their anal origins quite clearly . . . [in] some, for example, the whistling tic is derived directly from anal processes [flatus]. Here the patient carries out his hostile and degrading purposes by anal means." Over three hundred articles were written about the condition, each article going into the psychodynamic dimensions of the people suffering the disease and coming up with analyses not unlike those I've just reviewed. The theories were elaborate and, of course, logical, given the prior assumptions of psychoanalytical thinking.

There was a problem, however. Some neurochemists observed that the Tourette patients seemed to have low levels of a specific neurotransmitter, one of the chemicals that allow the brain's neurons to interact. An imbalance in these chemicals can give rise to bizarre electrical patterns of activity; apparently, bizarre behavior is not far behind. The scientists gave the patients the chemicals they lacked—at once their craziness was resolved.

Such data certainly neutralize psychiatrists' elaborate theories and explanatory constructs. And every year, researchers continue to close in on more realistic and purposeful explanations of behaviors that are frequently examined in their abnormal or ill state.

Affective Disorders

It is estimated that approximately 13 percent to 20 percent of the population has some depressive symptoms at any given time. The lifetime risk for major depression is between 8 and 12 percent for men, and between 20 and 26 percent for women. The prevalence of bipolar illness in both sexes was found to range from .65 percent to .50 percent in the industrialized nations. Depression has been observed in many different cultures, Western and non-Western. Throughout most of recorded history, there have existed literary and clinical descriptions of depression. Affective disorders are anything but new.

Freud believed that depression was not a symptom of organic dysfunction, but rather a massive defense reaction to which one resorts when one cannot cope with a stressful event—a view that leaves no room for selection theory. Although Freud realized that depression could have biological consequences, its ultimate cause was, in his view, purely psychological. Also, he differentiated between grief and depression. In grief, the reaction to the lost love object is conscious, while in depression, the reaction to the lost love object remains in the unconscious, and the ego is weak from repressing hidden feelings of sorrow and rage. Freud thought that depression could even be caused by an imagined or fantasized loss of a love object.

Other psychoanalytic theorists have argued that a person who experiences a withdrawal of love or gratification during the oral stage of development can become predisposed to depression. When a love object is lost, rage and reproach are directed inward rather than outward, and depression ensues. According to psychoanalytic theory, then, the therapist should bring the depressed patient to fuller consciousness of the stressful events in early childhood. Patients are encouraged to express their feelings of hostility rather than turn them inward. And, if patients can remember and work through their repressed childhood difficulties, they will be able to respond more realistically to their present situation.

The psychoanalytic point of view is similar to that of the "instructionists." According to the theory of "instruction," a person has no preexisting capabilities and is literally molded by the environment. Thus, everything the person knows or does can somehow be traced back to the environment.

Both the instruction and the psychoanalytic models would claim that affective disorders are caused by environmental events. Indeed, it is generally agreed that some depressions are caused by environmental events (that is, the reactive depressions). Some depressions, however, seem to occur without precipitating events: That is, a depression may spontaneously occur in a person who is not experiencing a stressful environment. Furthermore, affective disorders seem to be heritable, some people apparently having a genetic predisposition for developing such a disorder. Thus, environment cannot possibly be the only cause of affective disorders, and psychoanalysis cannot be completely accurate in its account.

Indeed, the affective disorders were described in purely psychological terms simply because there was no other way to describe them. We now know that major depression is a recurrent, heritable, syndromal illness that is accompanied by many observable biochemical changes. Modern brain science has provided significant insight into what goes wrong in the brain during depression, showing that purely psychodynamic processes do not somehow "cough-up" the disease. This psychological state is produced by disorders in normal brain chemistry—an apparently maladaptive trait to which I will return.

Recent studies have shown that the balance between two central brain neurotransmitters, noradrenline and acetylcholine, may play a role in the moderation of mood; and that depressive illness may be a disease of relative cholinergic predominance. Mania, on the other hand, could result from a relative adrenergic predominance. In addition, the hyperactivation of the hypothalamic-pituitary-adrenal axis is an important neuroendocrine characteristic of depression. It seems that there are increased levels of cortisol, ACTH, and beta-endorphin secretion in many depressed patients. Several investigators have suggested that acetylcholine plays an important role in ACTH and beta-endorphin release, which would further suggest that cholinergic neuroendocrine effects may parallel those occurring naturally in depression. Of course, however convincing the clinical evidence for these hypotheses may be, they are not likely to be the only explanations for the causes of affective disorders.

According to *DSM III-R,* a major depressive episode is marked by either depressed mood or loss of interest in all, or almost all, activities. Symptoms may include sleep disturbance, psychomotor agitation or retardation, appetite disturbance, change in weight, feelings of worthlessness, excessive or inappropriate guilt, difficulty in thinking or concentrating, and suicidal ideation. The average age of onset is the late twenties, but a major depressive episode may begin at any age. The onset of a major depressive episode is variable: In some cases, the symptoms develop over weeks; in other cases (as in "severe psychosocial stress") they may develop more suddenly. An untreated major depressive episode can last anywhere from six months to two years. An untreated depressed patient will experience a spontaneous remission of symptoms, and one is considered to be in full remission if one has had no symptoms for at least six months. In about 50 percent of the cases, the first major depressive episode is the last. But for the remaining 50 percent, the depression will return, perhaps many times. Episodes later in life tend to be more severe and begin more precipitately than earlier ones.

Much evidence supports the heritability of major depressive illness. Studies of twins show that the concordance rate for major depression is 65 percent among monozygotic twins, compared with only 14 percent among dizygotic twins. An adopted child whose natural parents have depression is two to three times more likely to suffer from depression than an adopted child whose natural parents did not have depression. Also, first-degree relatives of patients with unipolar depression have a higher incidence of this type of depression than the general population.

The heritability of major affective illness is further supported by recent studies done on the Older Order Amish pedigree. Researchers found certain sequences of DNA (called "restriction-fragment length polymorphisms") on chromosome 11 in subjects with unipolar, bipolar, and schizoaffective disorders, suggesting that these disorders may have a similar genetic basis. However, among those people in the Amish pedigree who have these DNA sequences on chromosome 11, major affective disorder has developed in only 63 percent. Clearly, the presence of this DNA sequence on chromosome 11 does not necessarily lead to affective illness, so perhaps predisposing environmental factors or other genetic factors are also required in order for the affective illness to develop. Also, while the concordance rate is higher among monozygotic twins than among dizygotic twins, it is still not 100 percent. Again, in addition to genetics, other factors must be involved in the development of affective disorders.

Whereas major depression is confined to depressive episodes, bipolar disorder involves both manic and depressive episodes. According to the *DSM III-R,* a manic episode consists of a "distinct period of abnormally and persistently elevated, expansive or irritable mood." During this period, there may be inflated self-esteem or grandiosity, decreased need for sleep, flight of ideas, distractibility, and reckless behavior. In bipolar disorder, the first episode is usually manic; it may be followed by a normal period, which is then followed by a depressive episode. The manic episode may also be followed immediately by a depressive episode. The mean age of onset of bipolar disorder is the early twenties. A manic episode may follow psychosocial stressors, antidepressant somatic treatment (drugs or ECT), and childbirth. Bipolar disorder is much less common than major depression. It occurs in both sexes with equal frequency, and the individual episodes are usually shorter and more frequent than the episodes of major depression.

Bipolar disorder has also been shown to run in families. First-degree relatives of a bipolar patient have a higher incidence of both bipolar and unipolar disorders than the general population. The concordance rate for bipolar disorder is 72 percent among monozygotic twins, compared with only 14 percent among dizygotic twins. Studies of the Older Order Amish pedigree have revealed a possible genetic location for bipolar disorder; but again, the results are not conclusive. However, there is some indication that genetic factors may be more important in bipolar disorder than in unipolar disorder. In first-degree relatives of subjects with bipolar illness, the frequency of not only bipolar but also unipolar illness is increased. It would be difficult to explain the biochemical basis for the "switch" of affect in bipolar illness if there were only one abnormality. Generally speaking, you would expect that some offspring of an individual with bipolar depression (that is, two abnormalities) might inherit only the predisposition to unipolar depression, whereas others might inherit both, hence exhibiting bipolar depression.

The original commonly held catecholamine (that is, epinephrine and norepinephrine) hypothesis stated that major depression is caused by a deficit of norepinephrine at critical effector sites in the central nervous system. In other words, decreased levels of brain norepinephrine were thought to lead to depressive symptoms, while increases of norepinephrine were thought to lead to manic symptoms. This hypothesis was supported by the fact that drugs that relieve depression increase the level of norepinephrine in the brain, while drugs that relieve mania reduce the level of norepinephrine in the brain. In addition, experimenters have measured a norepinephrine metabolite (MHPG) and found decreased levels of it in the plasma, urine, and cerebrospinal fluid of depressed patients.

Other studies show that serotonin and norepinephrine may both be involved in depression: L-tryptophan, an amino acid that increases serotonin levels, seems to be an effective treatment for both mania and depression. The serotonin-norepinephrine hypothesis states that a deficiency of serotonin creates a predisposition to affective illness; and that, given a serotonin deficiency, a high level of norepinephrine will produce mania, while a low level of norepinephrine will produce depression.

However, more recent studies that use direct assays of norepinephrine and mass-spectroscopic assays of the metabolite MHPG support a conclusion contrary to the original catecholamine hypothesis. These new studies report normal or increased levels of cerebrospinal norepinephrine, increased levels of plasma norepinephrine, and increased levels of urinary and cerebrospinal MHPG in depressed patients. Furthermore, successful responses to antidepressant medication are consistently associated with decreases in cerebrospinal and plasma MHPG in patients with major depression. The complexity of the neurotransmitter systems had made it extremely difficult to construct models of the specific role of neurotransmitters in depression. In addition, although there has been some success in detecting genes of major effect for enzymes involved in neurotransmitter synthesis and degradation, no one has produced compelling evidence for a major gene locus in affective disorders.

Now, one might well ask why our species has incorporated a capacity for depression. What adaptive value could it have for humans? Answering such a question is a dangerous game, but there are a few ideas on the topic.

The process of selection may be working at the personal level. For example, let's assume that X and Y are identical twins who were separated at birth, and both have gene(s) that predispose them to develop an affective disorder. We know they are predisposed through genetic screening techniques. Now, X experiences many stressful life events, but Y does not. Even though the twins possess the very same genetic predisposition, only X develops the depression. It could be postulated that X's environment has selected his preexisting capacity to develop affective disorders. Had we not known that X possessed this genetic predisposition for depression, we might conclude that the environment had "instructed" X to develop depression. But since we do know that X possessed a preexisting capacity for depression, it is more accurate to say that the environment "selected" that preexisting capacity.

Harry Harlow, the famous American psychologist who was among the first to hold out for biological constraints on behavior in the 1950s, at a time when behaviorism was the rule, found that, when separated from their mothers, infant monkeys displayed a separation reaction that was very similar to depression (fig. 8.2). Harlow has argued the following: The separation reaction "persists as a bio-behavioral defense against the disruption of the essential bond between the nurturing mother and infant." He found that the distressed separation reaction of an infant monkey elicited nurturing behavior by other adult females, and was adaptive for the infant monkey, since its physical and social needs would be fulfilled even though its mother was gone. Therefore, the depressive behavior may be seen as an important form of social communication.

Another explanation for how depression may be adaptive is found in the work of A. E. Schmale and G. L. Engel, who, in the mid-1970s, suggested that the "withdrawal following a loss, conserves the inner resources of the individual and permits subsequent adaptation to new environmental challenges and opportunities." Thus, the depressed or withdrawn organism is thought to be conserving resources and preparing for new challenges. This explanation applies especially well to people with winter depression (seasonal affective disorder), who may show more of the less common symptoms of depression: oversleeping, overeating, and weight gain. This disorder may be seen as adaptive because it involves strategies for economizing on energy when less food is available and ambient temperatures are lower.

It is also possible that major depression is not adaptive at all. One could argue that depression so severe as to be debilitating and, at the extreme, a precursor to suicide, is certainly not a valuable trait. Perhaps only mild forms of depression (the everyday "blues") have adaptive value and have been selected. In fact, it has been argued that moods have a self-regulatory function that serves to alert the organism to its emotional state. If this is the case, perhaps a blue mood is adaptive. But if mild depression is adaptive, how can we account for severe depression? Until a more satisfactory genetic explanation is found, it is unreasonable to assume that this disease is adaptive in any form, mild or severe. Still, the selection model, in taking into account both the genetic and the environmental factors in the illness, throws far more light on depression and other affective disorders than does the instruction model of psychoanalytic theory.

FIGURE 8.2 The famous monkeys of Harry Harlow. His laboratory studies were among the first to argue for biological models of affective mood states. Reprinted from Harry F. Harlow and Clara Mears, *The Human Model: Primate Perspectives* (Washington, D.C.: V. H. Winston, 1979), fig. 74.

We cannot realistically ignore the wealth of possible explanations bound to recent work in neurochemistry and brain genetics, nor deny the fact that environmental factors play important roles in the evolution of mental states such as depression. With this in mind, then, selection theory is a most attractive operational scheme.

PSYCHODYNAMIC PROCESSES

To disbelieve in the psychoanalysts' method of analysis of psychological states is not to deny the complexity of psychological states all humans experience. We are a marvelously complex bundle of desires, beliefs, loyalties, loves, hates, and so on. Our species thrives on these moods and thoughts. Surely something in our evolutionary experience can explain why we have accumulated all these mental states.

Back in the Pleistocene period, our capacity to register intentionally was beginning to take off. Since one's intention can easily be read in one's facial expressions, the capacity to mask one's intentions might well have evolutionary advantages. And, indeed, humans are uniquely equipped with a brain system both for managing spontaneous expressions and for setting facial expressions through voluntary and conscious control. The chimp, on the other hand, is incapable of making voluntary expressions. As David Premack observed, chimpanzees are capable only of making spontaneous facial expressions and, thus, have a hard time masking their intentions.

Humans have two neural systems for controlling facial expression (fig. 8.3). The system that controls voluntary expression is managed from the left hemisphere. It sends its messages to the nerve center at the base of the brain, which, in turn, sends out commands to the lower facial muscles, as follows: The left hemisphere sends a command directly to the right brain-stem nucleus, which orders the right lower facial muscles to respond. At the same time, the left hemisphere sends a command over the corpus callosum to the right half brain. The right half brain then sends the message down to the left facial nucleus, which, in turn, orders the left side of the face to respond. All of this is done so quickly in the normal brain that at the command to smile the face responds in a nicely symmetrical way.

Spontaneous facial expression is managed by a different neural pathway. First, unlike voluntary expressions, which only the left hemisphere can trigger, spontaneous expressions can be managed by either half brain.

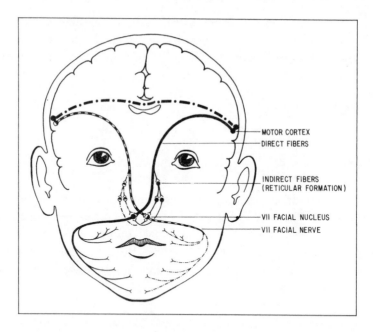

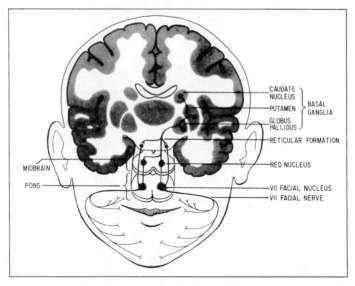

FIGURE 8.3 The neural pathways that control voluntary and spontaneous facial expressions are different. Voluntary expressions that can signal intention have their own new cortical networks in humans *(top)*. The neural network for spontaneous expressions *(bottom)* involve older brain circuits and appear to be common with those seen in the chimpanzee. Source: Michael S. Gazzaniga and C. S. Smylie, "Hemispheric Mechanisms Controlling Voluntary and Spontaneous Smiling," *Journal of Cognitive Neuroscience* 2 (1990): 239–45.

In this instance, when either brain triggers a spontaneous response, the pathways that activate the brain-stem nuclei are signaled through another pathway, one that does not course through the cortex. Each hemisphere sends signals straight down through the midbrain and out to the brain-stem nuclei. Clinical neurologists have known of the distinction between these two ways of controlling facial expressions for years. For example, a patient who has a lesion in the part of the right brain that participates in voluntary facial expressions will be unable to move the left half of the face when told to smile. At the same time, the very same patient is easily able to move the left half face when spontaneously smiling, since those pathways are unaffected by right brain damage. Also, as with Parkinson's disease, the pathways supporting spontaneous facial expressions do not work, whereas the pathways that support voluntary expressions do work. Such patients can lose their masked face appearance when told to smile (fig. 8.4).

It thus seems that the human has devised special brain circuits for masking intentions. One can voluntarily override, as it were, a spontaneous response to a stimulus. Once this mechanism is in place and becomes functionally active, we face what has been described as the "cognitive arms race." How clever can you become in masking intention?

Robert Trivers offers a brilliant idea on this topic, whose implications would find us right back on the doorstep of psychodynamic processes if it were not for the fact that he applies evolutionary theory to his assessment, not Freudian ideas about our murky subconscious. Trivers is fascinated with the manner in which we come to deceive ourselves. He reasons as follows: If caveman Jones is making voluntary expressions to fool caveman Smith, Smith should become increasingly sensitive (over evolutionary time) to the difference between real and fake expressions. Smith should be able to detect any twitch, seat, blush, quiver, pupil dilation, eye divergence, facial tension, or any other sign, no matter how slight, that indicates that Jones is lying. So what Jones must do (again over evolutionary time) is convince himself that what he says is actually true. That way, there is no duplicity, and the tension between voluntary and involuntary processes disappears. By having in your brain a system that enables you to believe what you express and do, there is no leaking out of affect, leaks caveman Smith could discern. Thus, there could be times when it might be adaptive to keep embarrassing truths from yourself (that is, from the area of the left hemisphere responsible for personal consciousness).

Are such psychodynamic processes going on in humans? Surely there are, and they are rampant. They are the result of the simple introduction

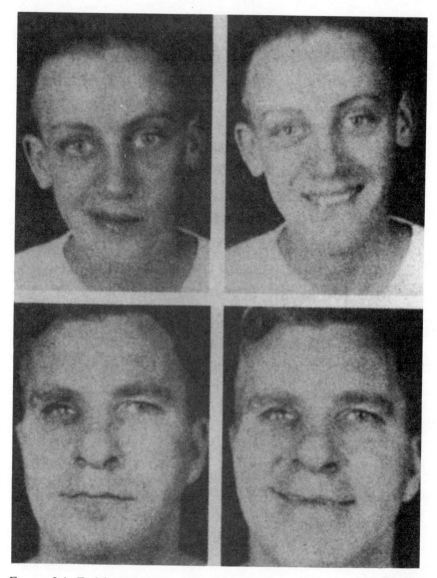

FIGURE 8.4 Facial expressions of two kinds of patient. The top patient suffered brain damage to the right hemisphere; the lesion interfered with voluntary facial expression. The lower panels show a Parkinson patient with a typical masked face. Although the diseases of both patients hit the part of the brain that controls spontaneous facial expressions, the faces of these patients, when told to smile, lighted up, since their other pathway was still intact. Source: R. N. DeJong, *The Neurologic Examination,* 4th ed. (New York: Harper & Row, 1979), figures 13.4 and 13.5.

of a special capacity for the human brain, the capacity to make voluntary expressions, which, in turn, allows for deception and the hiding of intentions. This capacity can, however, quickly compound, resulting in complex psychological states. From this viewpoint, selection theory gives us a rich way of thinking about how we humans have developed the complex psychological processes we commonly enjoy.

CHAPTER 9

Health Care, Aging, and Selection Theory

A common view is that while selection theory may be valid for much of human behavior, not all of our responses are inevitable; there are some aspects of our life that actually appear quite open to influence or even intervention from the environment. This, so the argument goes, seems to be particularly so in medicine. Medicine, after all, holds the keys to good health and longevity. If something goes awry in the body, medicine can provide a chemical, or irradiate, or—more simply—cut out the culprit, and the problem is gone. If that isn't instructing the body to fix itself, nothing is. Indeed, the instructionists' moment in the sun would seem unchallenged with respect to health issues.

One reason for this position is the fact that we completed most of our adaptations in the Stone Age—in an environment that was dramatically different from the one we know today. Our current diet, for example, is loaded with salt and fat, in concentrations much higher than our ancestors ever consumed; our air is contaminated with pollutants and other allergens; our social organization is different; and our visual environment now includes television and books. All of these factors have some effect on our health, and all seem to affect normal development and produce changes in our bodies. But do these environmental changes really render our Stone Age adaptations obsolete, or do they simply obscure the adaptive nature of the response?

Recently, evolutionary biologists have questioned the modern assumption that medicine holds the key to a long and healthy life. They see the modern human full of capacities to respond to disease and environmental challenge. In everything from morning sickness to psychological concerns such as "perceived control," the Darwinian trained scientist is beginning to wonder about how wise it is to try to intervene and override the body's own processes for dealing with the environment—processes that have been built in carefully over hundreds of thousands of years. Consider everyday normal pregnancy. Margie Profet, an evolutionary biologist, has shown how morning sickness is an adaptive response. It is during this phase of pregnancy that most women are lethargic, ill in the morning, generally don't like to eat much, and especially avoid substances with a bitter taste. Foods that taste bitter are reliable cues that the food may contain toxins. During the first trimester, plant toxins can have serious and negative effects on the formation of body organs of the fetus. Cutting down on food intake from a caloric point of view is not dangerous as the fetus is extremely small and requires little energy. Cutting down on toxins is very important. Evolution has built this in. Profet has shown that mothers who do not experience morning sickness have a higher incidence of miscarriage.

The point here is that responses that were selected for thousands of years ago still play an active role in our twentieth-century day-to-day life. In terms of what this means for health care, the question comes down to one, central issue: Should doctors treat patients in terms of twentieth-century intervention (that is, imposing technology on the body)? Or should they treat them in terms of adaptive responses that thousands of years of evolution have selected out (that is, guiding the processes that already exist in the body)? The latter option includes everything from understanding the long-term benefits of the body's responses, to how perceptions of control influence physiological functioning.

The famous evolutionary biologist G. C. Williams has provided an excellent perspective on this question, and is out to change our thinking about medicine, health, and the grand problem of longevity. In a 1991 issue of the *Quarterly Review of Biology*, he and Randolph Nesse, a psychiatrist, laid out the problems and issues in achieving their goal of physicians trained in evolutionary biology—trained to see, understand, and treat illnesses in terms of how the body naturally responds to disease and life-threatening events. This paper is sure to become a classic in the literature.

Williams and Nesse lay out the Darwinian approach for four major areas of medicine: infection, injuries and toxins, genetic factors, and the effects of abnormal environments. Overall, they argue for the use of the so-called "adaptationist program." Briefly, this program attempts to understand all biological phenomena in terms of their place in an adaptive biological system. Thus, a phenomenon can be seen either as a central and necessary component of the biological system or as an unavoidable cost to the system that comes along with an essential adaptation. Or, even more remotely, it can be some incidental manifestation of an essential operation. The assumption is that every part of the system has some adaptive role, and if that role can be defined, then it becomes possible to search for and describe other adaptive processes or structures related to this particular adaptation.

The adaptationist approach has been successful within animal biology. Williams and Nesse describe, for example, work done on the ampullae of Lorenzini, small organs on the top of the heads of sharks. For years, researchers were aware of the structure's presence, but had little idea what it was used for. Finally, it was determined that the organs respond to electrical currents; and in this way help the sharks to detect the heartbeats of potential prey. Additionally, a function such as the capacity of birds to navigate can also be examined in terms of its physical basis. In this case, *after* determining that some birds were responsive to the earth's magnetic fields, investigators found that these birds had magnetic particles buried in their skulls—hardly something one would look for without the functional knowledge. Similarly, one could approach medical phenomena in this light to determine the adaptive origin of particular physiological responses, and develop treatment programs that work with the body. Consider the following example.

Think of two humans walking along a rocky plain. One is a twentieth-century archaeologist surveying his next expedition, the other is the archaeologist's distant ancestor walking in the plains of Kenya a few hundred thousand years ago in the Pleistocene (better known as the Stone) Age. They both trip and sprain their ankles. In each case, there is immediate swelling and pain, and an increase in body temperature, leaving them both immobilized while their bodies rally their resources to repair the damage. Neither the archaeologist nor the Stone Age human "knows" what will happen, but evolutionary processes do, and that response doesn't make a distinction between a twentieth-century and a Stone Age ankle. As Williams and Nesse observe:

> Blood escapes from damaged tissues to cause a bruise. This and increased extravascular fluid contribute to local swelling. Histamine and other diffusible products of the injured tissues initiate the process that attracts phagocytes and other mobile cells, some of which start removing damaged structures and synthesizing their replacements.

180

An evolutionary biologist and adaptation-conscious physiologists and pathologists would ask a number of questions about the pathophysiology of the sprain. To what extent is swelling merely an incidental result of the trauma, and to what extent is it an adaptation to immobilize the joint or to otherwise favor healing? What harmful consequences may result from limiting swelling? What is the role of each cell type in the repair program and how are these roles coordinated for the efficient achievement of the repair? Are the repair processes influenced by temperature, and is healing fastest at a certain temperature? What exactly is the mechanism that results in pain, and is the pain adjusted to the expected need for immobilization under normal conditions of human ecology? Is local pain supplemented by more general injury-induced effects on motivation [lethargy and malaise]?

(Williams and Nesse, 1991, p. 9)

Although the adaptationist program has been used without question in biology to explain these physiological responses, the same process has not been applied to medicine. Williams and Nesse suggest that this is so because scientific medicine was born at the height of logical positivism "with its condemnation of all implications of purpose." This view is certainly ingrained in medicine, where most practitioners are interventionist by nature. Thus, physicians tend to approach such a problem from their own personal philosophy of care. Some (the interventionists) look only at short-term gain and comfort for the patient, and by reducing fever they may be prolonging the normal repair rate. For example, it is known that chicken pox is extended by a day if the patient takes aspirin during the course of the disease. Still other physicians would administer steroids to reduce inflamation—yet steroids are known to suppress the immune system, thereby increasing the chance for infection. Other physicians, however, would leave the sprain alone and tell their patient to get to bed. Which course is correct is up for grabs.

The importance of understanding the adaptive nature of our relationship with the environment is paramount when it comes to the issue of toxins. Every health fanatic wants a nontoxic environment, arguing that our Stone Age bodies were not prepared to deal with new inventions like pesticides and PCBs. They are right about Stone Age bodies, but they are wrong to think that a toxin-free world would be a good thing for our bodies. Williams and Nesse point out that plants defend themselves against being eaten by producing toxins. The chemicals that make coffee taste so good are actually toxins that prevent the bean from being devoured by insects and small mammals. The fact is, we have evolved certain detoxification mechanisms that allow us to consume coffee and other naturally occurring toxins safely. In addition, we prefer to eat a diverse diet—a practice that helps us avoid overloading on any one particular toxin, while ensuring that we obtain all of the necessary trace nutrients. Thus, while there is no diet that can be considered perfectly safe, we decrease the danger from toxins by eating a variety of different foods with very low doses of different toxins. Attempts to select artificially for disease-resistant plants would decrease pesticide use, but would concomitantly increase the concentrations of natural toxins in those plants, and a substantially increased concentration of even naturally occurring toxins would be detrimental to good health. The folks down at the health-food store do not think along these lines.

In addition, Margie Profet has recently shown that part of the allergic response is a protection against toxins. Substances that trigger allergies are usually either toxins or compounds that typically carry toxins. Allergy symptoms like vomiting, diarrhea, coughing, tearing, and sneezing all have the effect of ejecting the toxin from the body. Inflammation and lower blood pressure isolate the toxin, preventing it from hitting the bloodstream and being dispersed to the rest of the body, where it might do serious and immediate damage. The adaptationist would suggest that the usual prescription of an antihistamine to reduce the allergy symptoms might make for an increased risk to toxins, which could, in turn, increase the risk of cancer. Thus, what millions of years of careful groundwork has accomplished can be ruined by well-meaning intervention.

GENETIC DISEASES

On the surface, it seems to make no sense that Darwinian mechanisms would preserve genes that harm a species. Evolutionary biologists, however, distinguish between genes that create a low probable event and are part of an inbred line, such as Tay-Sachs disease; and diseases with a much higher probability of occurring, which are thought to have some adaptive value. Williams and Nesse's insight that "even diseases that severely depress both biological fitness and perceived well-being may be caused by genes with subtle beneficial effects, perhaps in other individuals or other stages of development," may explain why diseases like depression or schizophrenia survive in the biological system. Their expression in later life comes after the responsible genes have already participated in other, positive events during early development. More concretely, it has recently been reported that peptic ulcers, which are the result of genetic elevation of pepsinogen-1, are associated with higher proteolytic enzyme activity as the result of selection pressure to respond to the recent centuries of tuberculosis in our species. If that is true, popping antacids may well promote some bacterial infections.

Senescence is another disease that only recently has begun to be understood from the perspective of evolutionary biology. It was Peter Medawar, the great British immunologist, who first suggested that senescence is produced by genes that have beneficial effects early in life—when even slight positive effects of genes would be selected for, even if those same genes produce maladies in later life. Among other things, on the evolutionary scale, most animals do not live to old age; they are usually eaten at some point, thereby masking the effects of such genes. But in our industrialized society, modern science has conquered most of the events that used to cause early death.

We will face a difficult dilemma when genes that contribute to states like senescence are isolated and identified. The day is not far off when the process of genetic screening will be able to tell us whether a fetus has a gene for a particular disease. Yet, since the genes involved in senescence clearly have early beneficial effects on development, it would make no sense to block their action. The dynamic interactions within the whole organism are so complex as to make a simple fix impossible.

Another example of the deleterious effects of attempting to moderate genetic programs can be seen in modern-day efforts to control the body's metabolism, specifically in dieting practices designed to lose fat. Very simply, fat is an extremely efficient way for the body to stockpile energy, and the human metabolism is designed to store fat. On an evolutionary scale, the adaptive purpose of this was to ensure that energy reserves would be present during times when food from the environment was scarce.

This general principle applies to all humans, but the rate of storage varies considerably from person to person. At an extreme end of this spectrum, there is a class of individuals whom biologists classify as having "thrifty geneotypes." These people store calories in the form of fat very efficiently, but, because of modern living conditions, they never encounter episodes of famine to which they are especially adapted. As a result, these people are at risk to metabolic-related health problems, such as diabetes and obesity, and are urged to lose weight. There is a problem with this suggestion, though. Because of the adaptive basis of fat storage, attempts to restrict food intake voluntarily may be interpreted by the body as a period of uncertain food availability, resulting in adaptive inclinations toward food bingeing. This results in a vicious cycle of increases and decreases in weight, which may ultimately slow the metabolic rate even more. Such a response would be highly appropriate in a famine, but it directly opposes the point of voluntary dieting.

Finally, myopia has been examined. Clearly this is a maladaptive trait. Although myopia is a disease of modern societies, there is good evidence it has a strong genetic component. How can this be? Anyone born severely myopic in the Stone Age surely would not have survived.

But recent work has shown that eye growth is regulated by what the eye sees during development. Each eye grows independently, and we now know that the genes control how each eye responds to visual stimuli. If the eye is exposed to fine print or detailed information, it adapts and can lead to myopia. Thus, children at genetic risk for myopia might be well advised to read only books with large print.

THE RELATIONSHIP OF MIND AND BODY

Evolutionary thinking enriches our view of how our bodies and minds respond to disease and environmental signals. Also, as I have argued, our various mental states are to a large extent formed by selection pressures. Since, as we also now know, mental states can influence body functions, it is not surprising that particular mental states either enhance or impair our bodily response to environmental challenge.

The human need to control and influence the environment has been considered one of the great driving motivations behind our actions and behavior. When in control, we feel confident, capable, and assertive. In contrast, when deprived of control—when we are helpless before the forces of our environment—a sense of hopelessness pervades, and we become passive and withdrawn. Repeated experience with uncontrollable events can lead one to the conclusion that he or she is helpless and will be unable to control the outcome of any future events. Well documented in both human and animal research, "learned helplessness is characterized by a cluster of symptoms including motivational deficits, decreased persistence, cognitive deficits, and, in humans, sadness and lowered self-esteem (fig. 9.1). In addition to these psychological symptoms, a rapidly emerging body of literature has linked learned helplessness with a host of diseases including cancer, chronic heart disease, and arthritis, as well as chronic physical symptoms and diminished immune function. These findings may have important implications about the way in which we conduct our lives, perceive life events, and allow ourselves to be influenced by the environment.

	Learned Helplessness	Depression
SYMPTOMS	Passivity	Passivity
	Difficulty learning that responses produce relief	Negative cognitive set
	Dissipates in time	Time course
	Lack of aggression	Introjected hostility
	Weight loss, appetite loss, social and sexual deficits	Weight loss, appetite loss, social and sexual deficits
	Norepinephrine depletion and cholinergic activity	Norepinephrine depletion and cholinergic activity
	Ulcers and stress	Ulcers (?) and stress
		Feelings of helplessness
CAUSE	Learning that responding and reinforcement are independent	Belief that responding is useless
CURE	Directive therapy: forced exposure to responses that produce reinforcement	Recovery of belief that responding produces reinforcement
	Electroconvulsive shock	Electroconvulsive shock
	Time	Time
	Anticholinergics; norepinephrine stimulants (?)	Norepinephrine stimulants; anticholinergics (?)
PREVENTION	Immunization by mastery over reinforcement	(?)

FIGURE 9.1 The phenomenon of learned helplessness. From Martin E. P. Seligman, *Helplessness* (New York: W. H. Freeman, 1975). Copyright © 1975 by Seligman. Reprinted with permission of W. H. Freeman and Company.

The extent to which you feel in control of your environment may not only help determine your mental state, but it may also predict your physical health and longevity as well. Clearly, the brain mechanism that allows for the "feeling of control" is one that has considerable survival power and thus must have been selected for in evolution. With this in mind, society must examine the social programs, policies, and institutions that, though established to provide and care for us, actually minimize our influence over our environment: That is, such programs as welfare and institutions for the elderly, may actually, in disempowering the individual, be detrimental to one's mental and physical well-being.

Psychoneuroimmunology—the study of the interaction of the mind, the nervous system, and the immune system—has generated an abundance of evidence supporting the existence of physiological links between the state of the mind and the state of the body. The neuroendocrine system, which releases hormones in response to the activity of the nervous system, is directly influenced by our cognitive state. The immune system protects us against infection and destroys unwanted substances in our body. The physiological commonalities and influences these two systems have on one another are becoming increasingly well understood by psychoneuroimmunologists. For instance, the immune system has cells that are receptive to hormones produced by the neuroendocrine system, and the neuroendocrine system has cells that are receptive to hormones produced by the immune system (fig. 9.2). Lesion studies have also suggested physiological connections between the two systems. For example, lesions to the anterior hypothalamus, a brain area highly active in neuroendocrine processes, were found to inhibit lymphocyte stimulation (an immune response) in the guinea pig. Also, stressful activities have been found to increase catecholamine levels (that is, epinephrine and norepinephrine), which have been associated with hypertension and chronic heart disease. Thus, what we think or how we perceive the world may ultimately influence our immune system's ability to ward off disease via the neuroendocrine system. These findings help to provide a physiological explanation of how mental state can affect physical health.

A crucial point is that stress itself has not been shown to be detrimental. Rather, a perceived sense of *uncontrollable stress* has been associated with physical symptoms and illness. Some of the most compelling research to demonstrate the comparative effects of controllable and uncontrollable stress have come from paradigms resembling the learned helplessness experiments that produced listless and depressed animals. In a classic experiment, one set of rats was exposed to a series of shocks that were escapable by pressing a foot bar. Another group received an equiva-

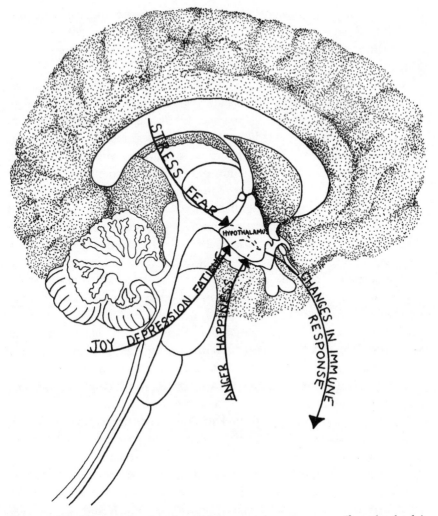

FIGURE 9.2 The mechanism by which mental events can affect the body's capacity to marshal an immune response. Mental states play on the hypothalamus, which in turn modulates the immune system. Adapted from Arthur C. Guyton, *Basic Neuroscience: Anatomy and Physiology* (Philadelphia: W. B. Saunders, 1987), p. 11.

lent amount of shocks, but their shocks were inescapable. A third group of rats did not receive any shocks. All of the rats were then injected with mitogens, which normally invoke an immune response. The rats' lymphocyte proliferation was then analyzed as a measure of their immunological functioning. The results showed that only the rats receiving inescapable shock had suppressed lymphocyte production. The rats that could control the shocks had an immune response equivalent to those of the rats receiving no shocks at all. Thus, it is not the stressful event itself, but rather the animal's perceived control of the event that can alter immune function.

Related animal research using controllable and uncontrollable stressors has yielded similar results. For instance, loud noises have been found to suppress immune functioning in mice. Infant monkeys separated from their mothers had suppressed lymphocyte proliferation in comparison to a control group. When the monkeys were reunited with their mothers, lymphocyte proliferation returned to normal. Besides demonstrating that an uncontrollable stressor such as maternal separation can lower immune function, this experiment is significant because it also demonstrates that alterations to immune function need not be permanent. By changing the environment, and thus the individual's cognitive state, proper immune function can be restored.

Here again is a response that may have been selected for in the Stone Age, when it is likely that children played outside the habitat and were always close to a parent. Their attachment was strong, much stronger than contemporary societies might like to admit. As Williams and Nesse point out, there is a burgeoning literature on the effects of extended day care by strangers which suggests that, in the main, it is more stressful to a child than being with a parent and therefore tends to reduce the immune capacity.

Human experiments performed in the laboratory concur with the animal literature. Natural-killer (NK) cell activity, a function of the immune system, has been shown to be significantly reduced by examination stress, loneliness, and other environmental stressors. For example, one experiment was conducted in which healthy volunteers were exposed to a series of loud noises (100 decibels) under controllable and uncontrollable conditions. Subjects in the uncontrollable condition reported increased feelings of helplessness and anxiety as compared with the control condition. More important, the uncontrollable-noise condition resulted in altered neuroendocrine and autonomic nervous system functioning, whereas the controllable-noise condition did not. Thus, humans appear to be subject to the same negative consequences of uncontrollable stress as rats.

Even more compelling research has been conducted outside the laboratory in real-world settings. In these studies, feelings of control over stressful life events have been correlated not only with general happiness but also with physical health. For instance, a longitudinal study of personality and stress resistance reported that feelings of self-confidence and the tendency to confront rather than avoid stressful situations served to protect individuals from the negative psychological consequences of life stress, and, in a study of women, even minimized psychosomatic complaints such as headache, acid stomach, and insomnia. The term *hardiness* has been used to describe those persons most resistant to stress. In a group of business executives, hardiness was found to be characteristic of executives who remained healthy when under stress, as opposed to those who became sick. Hardiness and a sense of control have also been associated with better physical and mental adjustment to cancer.

Conversely, people experiencing uncontrollable stressful events, or people considered to be poor at coping in general, tend to experience more symptoms of helplessness and decreased immune function. In a longitudinal study of Harvard graduates, investigators identified pessimism in early adulthood as a risk factor for poor health in middle and late adulthood. Studies of populations exposed to chronic or acute uncontrollable stress also indicate that these populations experience lowered immune functioning. For example, caregivers and family members of patients with Alzheimer's disease were found to have increased stress-related psychiatric symptoms, poorer immune function, and increased incidence of depression. Immunological suppression was also found during conjugal bereavement, lasting anywhere from four to fourteen months after the death of a spouse. This finding may explain the phenomenon of increased mortality rates for widows and widowers. Depression, in general, has also been correlated with immunologic suppression in comparison with other psychiatric disorders. Thus, in accordance with the experimental data, an accumulating number of field studies have documented that lowered immune function can and does result from uncontrollable stress in our ecological environment.

While there are many measurable effects, there are, of course, also limits to how much perceived control can accomplish. Take the stark statistic of mammalian life that males tend to die before females (fig. 9.3): Males carry more, and usually more severe, infection; males die more frequently from accidents; males are murdered more often; and males die more frequently in utero. Why is this so? Certainly, males are accused of being more aggressive than females and are generally perceived to be more in charge. Why are males not spared from their dismal rate of mortality relative to females?

The favorite hypothesis for years was that the culprit was the males' unguarded X chromosome. Yet genetic explanations of this kind seem doubtful. As Robert Trivers has reviewed in his *Social Evolution* (1985), the phenomenon is most likely related to the sexual differentiation dependent

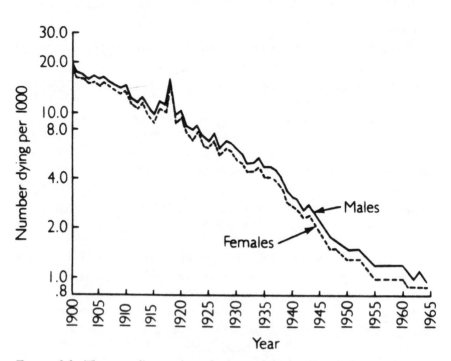

FIGURE 9.3 The mortality rate in males is much higher than in females. Source: National Center for Health Statistics, *Vital Statistics of the United States* annual volumes.

on the gonads. For example, in Kansas at the turn of the century, a number of mentally retarded males were castrated. When the mortality rates of these men were examined, they were lower than those of noncastrated men, and some of these men even outlived many females. Also, it was noted in these studies that intact males had a higher incidence of infection. In short, by throwing the longevity issue into endocrine mechanisms, one can begin to see how the perceived control interacts with this kind of system while also being limited in the overall extent of its action. Consider recent work on brain aging.

It has been suggested for some time that stress can affect the aging process. While the biological system that supports the stress response is a needed and welcome partner in our body's defense against outside unwanted challenges, it can be somewhat of a double-edged sword, and impair good health as well. A key chemical player in the stress response, the glucocorticoids, helps to marshal our heart, muscles, and other systems to ward off physical threats quickly and effectively. Reallocation of our body's energy resources to carry out these emergencies postpones other normal bodily functions, such as immunosurveillance, until the stress situation is over. At the same time, the chemicals secreted to allow for this quick, energetic response may linger a while after the stress response, and, ironically, their presence in more normal body states can be harmful to muscles, induce hypertension, and produce a host of other problems. In short, there are no free lunches. The stress reaction is both good and bad.

It has also been suggested that stress can actually affect the brain and perhaps even promote the death of our brain's neurons. Indeed, connecting stress mechanisms with the brain was an exciting insight into the relationship between the mind and physical health. The traditional view is that the brain's cortex influences a small structure at the base of the brain, the hypothalamus, which is important in regulating bodily functions. This structure, in turn, directs our pituitary gland to trigger the adrenal cortex, existing in our adrenal glands, to secrete the glucocorticoids, which are active in the stress response. This brain-to-body connection is the reason many people now believe that our mental state can influence systemic bodily functions and thereby influence our state of health.

Recently Dr. Robert Sapolsky of Stanford University questioned whether those same busy chemicals, the glucocorticoids, are capable of attacking and killing brain cells under some conditions. Sapolsky pursued an earlier finding of Dr. Bruce McEwen of Rockefeller University, who had shown that an important brain structure, the hippocampus, was full of cells that were especially responsive to these chemicals. This finding in itself has raised interesting possibilities for the understanding of human mentation, especially memory phenomena. Brain scientists have known for years that damage or surgical removal of the hippocampus severely impairs the ability to store new memories. What might happen to memory if this structure became overexcited? The chemicals released from a stress response would activate this brain structure—a possibility that might become exceptionally good following a sudden stressing event. It has been suggested that this might be the case and might explain why people tend to have extremely vivid memories of events surrounding a sudden, stressful event.

But alas, Sapolsky's results, which force us to consider the down side of the stress response for this very same structure, point to a different aspect of hippocampal function. Using rats, he discovered that extensive exposure to glucocorticoids promotes cell death in the hippocampus. He has also shown that rats become impaired at solving spatial memory problems, such as remembering where they are in relation to a food reward, whereas we remember where the kitchen is, no matter what room of the house we are in. Sapolsky proposes a sort of cascade effect that is triggered by stress, which eventually leads to diminished mental function, such as spatial memory. In his view, following stress, cell death is fostered in the hippocampus, which, in turn, disrupts its normal functioning and, ultimately, its ability to assist with memory functions.

Even more intriguing is Sapolsky's observation concerning rats that are handled while young. In this laboratory version of tender loving care, it was found that these rats remained somewhat insulated from the cascade effect into accelerated dementia. Rats that are ignored and left alone in their home cages, merely to be routinely fed and cared for, are more likely candidates for the downward spiral of brain aging.

Sapolsky and his colleagues reported for the first time that such processes may be going on in an animal with closer ties to humans—the vervet monkey. He examined eight such monkeys that had died spontaneously while housed in a private center in Kenya. All had multiple gastric ulcers, which suggested that they had experienced stress during their lives. Examination of the brain revealed marked degeneration in the hippocampus. Brains of other monkeys that had been euthanized for other research reasons showed no such damage. The investigators also noted that the monkeys that had the hippocampus damage also had a high incidence of bite wounds. This suggests they were under much more social stress and were most likely part of a subordinate social group within the monkey colony.

The bleak consequences of stress for rat and monkey are both intriguing and alarming. Is there evidence that this kind of death march is occurring in the human brain? At this writing, the apparent conclusion is that the stress system continues to work normally as we age. The dynamic feedback system that turns the stress chemical response system on and off appears to work well late into life. However, there are indications that the cascade effect is present in humans but hidden—except in the presence of other diseases, such as Alzheimer's, when hypersecretion of glucocorticoids may further accelerate the deterioration of the aging process.

Sapolsky offers us a fascinating picture. The overproduction of glucocorticoids is known to have a disastrous effect in some animal species, most notably in some species of marsupial. After an intense mating season, the male dies as the result of a massive oversecretion of glucocorticoids. This programmed death occurs in other species and is always tracked to the overproduction of the multifarious glucocorticoids. These are extreme cases, and Sapolsky is quick to point out that the reactions of our species to glucocorticoids are more attenuated. Nonetheless, the relationships are there; and in the years to come, more research ought to spell out the dangers of stress for our own brain health during aging.

SOCIAL DEPENDENCY AND PERCEIVED CONTROL

In light of the overall empirical findings, what is the value of government-sponsored social programs designed to help us by taking care of all of our day-to-day worries? Certainly the intentions are good: Social support systems, welfare, and retirement homes all make life a little easier by solving the complex situations that may arise in a lifetime. However, in doing so, these institutions also remove the feelings of personal responsibility, efficacy, and control that one needs if one is to act and make decisions affecting the immediate environment. An individual who cannot make an impact on his or her environment and exercise personal choice will never feel the sense of control and self-sufficiency that has been demonstrated to be vital to physical and mental health.

A good deal has been written about the importance of a social support network in maintaining health and personal well-being. Social support, which can take the form of a spouse, family, or friends who provide psychological and material resources, has been correlated with increased health and happiness. Individuals with social support networks have been found to have decreased incidence of illness and lower mortality rates than do comparable populations who live in relative social isolation. It is hypothesized that these support systems are beneficial because they may both deter harmful behaviors (like smoking and drinking) and be a source of health care, food, housing, and other necessities. Psychologically, individuals with strong support networks have been found to have higher self-esteem, greater self-identity, and an increased sense of control over the environment. These positive feelings exist largely because people feel confident that, with their social support network, they have the capability to handle any stressful event that might arise.

Although a social support network is clearly of immediate use, what happens when it is no longer available? If a social program established explicitly to provide a support network becomes inaccessible, the members may be left in a state of dependency worse than if they had received no help at all. By creating a false sense of competence, social support networks may ultimately leave people ill equipped to cope with and control their environments.

Numerous longitudinal studies have confirmed this disturbing hypothesis. Most noteworthy is the Cambridge-Somerville Youth Study implemented in 1939, which was intended to be a controlled study advocating social support in an effort to prevent delinquency. The participants, seven hundred boys from urban areas of Massachusetts, were carefully matched and then randomly divided into a treatment group and a control group. Each boy in the treatment group was assigned a counselor who visited, on average, twice a month. The boys were encouraged to use the program for family assistance and academic tutoring. Medical and psychiatric attention was given to those in need. Most were introduced to the YMCA, Boy Scouts, summer camp, and other community programs. In contrast, youths in the control group were merely required to supply information about themselves and received no counseling or social support. Some thirty years later, over five hundred of the participants in the study were located by searching court records, hospital records, and city directories. The control and treatment groups were compared in terms of health, family, work, attitudes, and criminal behavior.

The results are astounding. The treatment group had a significantly higher incidence of stress-related disease such as arthritis, emphysema, depression, ulcers, high blood pressure, and other circulatory system ailments. A higher percentage of the treatment group displayed symptoms of alcoholism and/or signs of serious mental illness. Although an equal number of men in each group had died by the time of the follow-up study (twenty-four in each group), the members of the treatment group had died at younger ages. The only difference in criminal activity was that a significantly higher proportion of treatment-group criminals committed more than one crime in comparison with control-group criminals. The control group was found to have more white-collar workers and professionals than the treatment group, and the control group had greater job satisfaction.

What can we conclude about the value of such a "treatment" program? Subjectively, many of the respondents who had been in the treatment group rated the program highly and expressed gratitude for the help and support they had received. But objectively, the results cannot be denied: Those individuals who received no help or counseling and were left to their own devices were markedly better off both physically and emotionally in the long run. The author of the follow-up study, Joan McCord, has speculated on why the program may have actually harmed the treatment group:

Agency intervention may create dependency upon outside assistance. When this assistance is no longer available, the individual may experience symptoms of dependency and resentment. The treatment program may have generated such high expectations that subsequent experiences tend to produce symptoms of deprivation. Or finally, through receiving the services of a "welfare project," those in the treatment program may have justified the help they received by perceiving themselves as requiring help.

(McCord, 1978, p. 288)

Related studies have drawn similar conclusions that "treatment programs" intended to prevent crime and delinquency may not only be ineffective but may actually have had negative consequences for the participants.

These failed programs highlight a major flaw in many of our existing social programs and institutions. For example, in nursing and retirement homes, each resident is unconditionally provided with all of his or her daily needs—food, medication, clean sheets, and a bath. Although this may sound reasonable and perhaps even enjoyable, nursing homes have been found to increase dependency and contribute to depression and social isolation among the aged. Direct observation of nursing homes has found that the staff reinforce dependent behavior and either punish or do not respond to independent behavior. By encouraging passive, dependent behavior, such institutions inhibit the elderly's ability to exercise personal choice and exert control over their environment, which, in turn, may be detrimental to their physical and mental well-being.

Likewise, in our nation's welfare programs, a correlation has been found between low socioeconomic status, low sense of control, and poor health. By providing for the needy, without requiring them to earn what they receive, we diminish their perceptions of self-worth and increase their feelings of helplessness. Our current welfare system not only does nothing to promote a person's sense of being able to take charge of his or her environment, but it also reinforces the conviction that one is in need of help, and thus leads to depression and physical illness as well. Only a person who is convinced of his or her ability to exert control over the immediate environment will be able to rise above the abject conditions of poverty and function autonomously in society.

Several real-life experiments using "perceived control" as the independent variable point to the significant benefits of efforts to increase a sense of control in populations. Judy Rodin and her colleagues at Yale University conducted an experiment suggesting the benefits of enhanced personal responsibility and choice among nursing-home residents. Differentiated only by which floor of the nursing home they lived on, one group of residents, identified as the comparison group, was given little personal choice in the daily routine and activities. It was impressed upon these residents that the staff was responsible for caring for residents and for meeting all their needs. The second group of residents, the intervention group, was given many choices and told that it was the responsibility of every individual to make the nursing home what he or she wanted. Each person in each group was also given a plant: The staff took care of the first group's plants, while the second group had to take care of their own. After eighteen months, residents in the intervention group were found to be happier and more alert, with fewer complaints of physical symptoms. Even more compelling, the mortality rate for the intervention group over the eighteen-month period was 15 percent for the intervention group and 30 percent for the comparison group. (The previous mortality rate for the nursing home was 25 percent overall.) Even the experimenters were amazed at the results achieved by using only such subtle variables as telling people to make their environments what they wanted it to be and giving them plants to take care of. Imagine what more direct approaches could do! A joint task force for the American Medical and Nursing Associations has concluded, "A sense of purpose and control over one's life is integral to the health of the aged."

Of equal importance is the way doctors encourage patients to cope with illness. Although doctors may prefer to work with the compliant and accepting patient, it has long been known that the externally angry and uncompromising cancer patient has a statistically higher survival rate than the passive equivalent. Patients who resist a state of helplessness and express a belief in control over their cancer tend to be better adjusted overall. Perhaps physicians should not encourage compliance and passivity, but rather should channel the energy of a patient's anger toward a willingness to fight and control the cancer. A sense of helplessness can only weaken the body's ability to defend itself against the disease.

Also in the realm of medicine, many self-help and self-regulation treatments have proven more effective than administration under medical supervision. One study of patients with rheumatoid arthritis concluded that increased control over one's medical care and treatment was correlated with better mood, psychosocial adjustment, and fewer daily symptoms. Similarly, patients requiring hemodialysis reported more control over dialysis and less pain when they were able to self-direct their treatment as opposed to having staff-managed treatment.

Programs encouraging personal choice and control have been reported to be successful in nearly every aspect of daily living. In the classroom, increased choice and the ability to make decisions has been shown to improve attendance and enhance the intrinsic motivation to learn. On the job, allowing workers to play a role in decision making through participative management has increased organizational effectiveness, improved attendance, and increased job satisfaction.

Strategies to maximize personal control can be implemented in nearly every aspect of daily living—not only for the poor, the elderly, and the sick, but for the housewife, the worker, and the student. A learned helplessness state is exactly that—a *learned* phenomenon. With appropriate intervention and treatment strategies, one can learn both that one is not helpless in the environment, and that one can acquire the ability to make personal choices and exercise the control that has proven to be vital to well-being. The evidence is clear: We stand to maximize our productivity, promote physical health, and enjoy personal happiness if we can learn to take charge of our environment. This is true for all members of our species. Selection processes got us here and gave us the ability to develop beliefs and hypotheses about the world. Armed with the knowledge that our species has that capacity, and that taking control of the environment is beneficial, we can become full-fledged actors in our lives rather than being dependent on our environment.

Selection Theory and Clues to the Nature of Conscious Experience

I
N the preceding chapters I have spelled out the nuts and bolts of selection theory as well as its implications for some of the social, medical, and scientific questions of our day. It is a theory about brain and biological function that I find satisfying and comforting at many levels. It not only makes sense of a tremendous amount of seemingly conflicting information, but it allows for some musings about the nature of human conscious experience itself.

Over the years there has been an inevitable tendency on the part of the critics to weaken selection theory by dismissing it as an aspect of what is innate. That way of thinking misses the point of the power of selection theory—its ability to account for the confounding variability of behavior. We all know that different animals respond differently to the same environmental challenge, and that animals of the same or similar genotype respond differently to varying environmental challenges. It is these kinds of observations that have led students of behavior to believe that the organism is ready and able to adapt; that it can receive instruction from the environment; and that it can become modified in some way and then store that information for the next time the same challenge presents itself.

What I have argued is that selection theory takes the same information and points to another model. The strong form of the argument is that an organism comes delivered to this world with all the complexity it will ever have already built in to it. In the case of the modern brain, a range of circuits that enable a variety of behavioral and cognitive strategies become matched with an environmental challenge and the selection process starts. What looks to be learning is in fact the organism searching through its library of circuits and accompanying strategies that will best allow it to respond to the challenge. When the concept, which is well established in basic biology, is applied to more integrative mechanisms of mind, it goes a long way in explaining not only complex phenomena such as learning, language acquisition, mental disease, and variations in intelligence, but also everyday patterns.

Selection theory makes sense out of many things—even a walk-through of the medical establishment of China. In our culture there are well-established behavioral patterns associated with the various medical specialities. The surgeon is commonly characterized as dominant, brusque, and assertive. The internist is circumspect, passive, and more thoughtful. The radiologist is entrepreneurial, frantic, and fast-talking. Each of these stereotypes has been explained away in our culture as reflecting economic issues that so dominate their trade. The surgeons make all the money and as a consequence move independently of the checks and balances of the lower-paying specialties. The hustling radiologists are paid by the piece and are constantly trying to increase the number of scans they can read for a fee. The internists are by and large working for set fees. They can only charge so much, and their work takes time. Until recently, they had few procedures to run and as a consequence were not that wealthy. Well, in China almost everybody gets paid the same amount. What are the doctors like? The answer is that the personality types are virtually the same as their American counterparts. In short there seems to be something about the specialty per se that selects the people who go into it. It is as if there were particular traits in a personality that draw one to particular ways of spending their lives. The economic hypothesis is not valid. Ideally, as we come to a deeper understanding of why people do what they do, we will have a far richer, more biologically based, theory of personality.

Within our own culture, I think we all muse about the differences we note about the various disciplines around us. When I was at Caltech there was a story about Richard Feynman, the great theoretical physicist. He decided one year to spend a sabbatical with the equally great Caltech biologist Max Delbruck. Feynman was in the lab every day for a year, trying to make a go of it in the new molecular biology. He finished some experiments but at the end of the year announced he was going to leave biology to the biologists, that he simply didn't think like they did. It is not clear whether such selections are based on motivational or cognitive dimensions of mind. It is clear, however, that they go on all the time. My guess is that those drawn to formal philosophy have special systems that the rest of us do not possess. I really don't think Wittgenstein happened to be a philosopher.

Second, the knowledge that we now have about how the young mind develops almost dictates a selectionist view of the process. The notion that the vertebrate brain comes to the world naive and that experience writes its story on a clean slate is one of those marvelously simple ideas that is simply wrong. Modern developmental psychology has grown with the philosophical insights of Quine and others who point out the importance of a priori constraints—constraints that biology has shown to have been built up through millions of years of evolution. The fact is, there is only a limited number of interpretations that we can possibly apply to a given set of data. The implications for the developing mind are strong about this point, as is evident from the data on children's limited ability to name objects in the presence of a label, the apparently innate limitations of the perceptual system, and so on.

Parents who have thought about it know that they don't have much influence over their children. A parent who tries to instruct his or her child often becomes frustrated and, as a result, can inadvertently create an environment that can do a lot of harm. A parent's job, society's job, is to set a context for opportunity. Each of us has our individual talents and set of limitations. The trick is to match those up in a positive way. It seems so simple, yet it is very hard.

Another limitation we are forced to recognize from selection theory is the precariously set nature of our adult neural networks. Everybody would like to believe the brain is infinitely adaptable and almost impervious to injury and disease. We all want to believe that the right set of instructions from the environment will remake the underlying tissue and allow for normative behavior. Every institution in our country is given over to this happy thought. Indeed, even within the hollowed walls of science, the belief is in brain plasticity and repair. We are a well-meaning species and we want to be able to fix things that are broken, especially if we think that we broke them. Yet selection works on the substrate—the delivered brain. Attempts at instruction processes also work on this same substrate, and yet the brain seems strangely unyielding. While modern neurobiology has demonstrated its ability to create conditions where neurons in the central nervous system grow, no one has shown that any of this activity is of any functional consequence. Indeed, one can argue unplanned growth is detrimental to the brain. Take the macabre example of serious closed head injury.

Young males have the highest incidence of severe head trauma, and depending on the initial severity of the injury, there is a fairly bleak outcome. The adult brain is damaged through trauma at both the cellular and network level. The patient is usually placed in a rehabilitation hospital with the hope that the damaged tissue will repair itself. Neuroscientists seeking funding hold out the possibility of meaningful neuronal repair. Clinicians tend to point to a particular case in which the patient did get a lot better. If significant repair were a possibility, however, it ought to be possible in the young adult brain, which still holds some promise for plasticity. The selection process that is normally at work in learning, whereby strategies are sorted through until the one appears that can be useful in coping with an environmental challenge, is still at work. Following the injury, however, the number of available circuits is down and others don't seem to be taking their place.

A similar situation can be seen with the premature baby. It was considered a great medical success a few years back when physicians were able to save babies who were significantly premature at birth. Now it is routine. However, we now know that their brains are not normal. Their brains are loaded with abnormal connections and aberrant synaptic connections. We also now know that children who were born prematurely do not perform within the normal range of achievement. Selection is at work during their life, but, in these cases, the process has less to draw upon.

Selection theory does have its comforting aspects, however. Parents whose child becomes depressed or schizophrenic, or has what they might consider a bizarre sexual preference, or cannot solve Maxwell's equations, or is obstinate or headstrong, or whatever, need not berate themselves. Extreme environmentalism is cruel because it suggests to the parent that their child has been warped by something they did. The evidence points in the other direction. Theories like Bettelheim's "icebox mother" theory of why some children suffer from autism and Laing's "double bind" theory of schizophrenia are all nonsense ideas that blame genetic deficits on dear old mother. The purpose of parenting is to provide steady and wide-ranging opportunities for a child. The child does the rest. It is as simple as that.

Fourth, and most important, the ideas put forth in this book allow for a fresh look at the nature of human conscious experience. Considering all of the domain-specific capacities of the human, capacities that have been carefully built into our brains over millions of years of evolution, some might have a denigrating view of who we are and what we accomplish. Others want the sensations we experience as human conscious agents to be the product of the vast computational power our huge cerebral cortex must allow—with each person starting from scratch and building up their own story over a lifetime. This mix of ideas generates the unclearly stated and fairly vague view that human conscious experience is a "thing" that emerges from the human being's information-processing capacity. Well-meaning scientists, philosophers, and YMCA leaders all write books about it. Each one that is read, makes sense until the next one is read. The experience of reading these books is reminiscent of the novice reading lawyer's briefs: The first brief seems to make a really good point until the next one is read. It seems to make a good point too. Then back to the rejoinder, followed by the rejoinder, and so on.

The modern human is a bundle of special-purpose systems that allow us to communicate, evaluate facial expressions, make inferences, interpret feelings, moods, behaviors, and all the rest. Studying patients with brain damage reveals how specific these capacities can be. One capacity can be removed without necessarily affecting another. Each of these activities is managed by neurons in ways that no scientist fully comprehends. Millions upon millions of neural impulses are churning away to produce human talents. Yet to say that that is consciousness is somehow missing the point. I am beginning to think there is another way of viewing these issues, building on the realization that human consciousness, at its core, is a feeling—a feeling about special capacities. My guess is that we are looking in the wrong wood pile for the answer to the problem of human consciousness.

There are some obvious aspects of human consciousness that we tend to lose sight of. First, one does not learn to be conscious. When the brain starts to function, up it comes, just like steam out of a turbine. There is no getting rid of it. And the feeling of consciousness is not unlike other seemingly unfathomable feelings like the feeling or instinct to survive. It is there. Oddly, most philosophers and biologists have not tortured themselves about understanding that feeling. Yet it is surely no simpler to understand and know the mechanisms about then human consciousness itself.

Second, the feeling of being conscious never changes in life. A few years ago my seventy-six-year-old father, a physician of enormous intelligence and savvy, would sit in his easy chair contemplating something or another after years of strokes that were slowly consuming his cerebral cortex. Mercifully, the strokes did not impair his language and thought processes. He knew of my profession and the kinds of issues that interested me. I asked him how he felt, and he replied, quite simply, "Mike, I feel twelve. I always have and I always will." For him, his consciousness was the same. The computational skills were vanishing just like they do for all aging brains, but the feeling of being conscious never seems to disappear.

In my own aging process I find it almost hilarious to look in the mirror. Looking back at me is a fifty-year-old man. Looking into the mirror is a person who also feels twelve. Even though my cerebral cortex and my body have undergone vast changes from the age of forty on, I feel the same as I always have. To put a little neural meat on this idea, modern brain science knows that subcortical structures are heavily involved in the management of feelings—of felt states. Many of these systems change very little with aging. They stay the same and it is these brain areas that pump out the chemicals and activate their circuits for generating the feelings associated with the specialized perceptual and cognitive capacities we humans have accumulated over millions of years. Consciousness is a feeling, a feeling about things that doesn't seem to change.

Selection theory points us in the direction of specialized systems. It seems to me these capacities are always associated with feelings. A six-hundred-dollar Japanese video camera may be able to see the world, but it can't *feel* the world. Our subsystems in our visual system, intricately evolved, allow us to see the world, and associative feelings that we have about those images make us want to see, and to see more. Without that feeling about seeing, about talking about anything, our desire to survive, to reproduce, to share this sensation with our offspring would be absent.

Suggested Readings

1. A Lesson from Biology

A number of texts provide a more detailed understanding of immunology and biological processes, including J. T. Barrett, *Textbook of Immunology: An Introduction to Immunochemistry and Immunobiology* 5th ed. (St. Louis: C. V. Mosby, 1988); H. B. Barlow and J. D. Mollon, eds., *The Senses* (Cambridge: Cambridge University Press, 1982); S. Luria, S. J. Gould, and S. Singer, *A View of Life* (Menlo Park, Calif.: Benjamin/Cummings, 1981); J. E. Cushing and D. H. Campbell, *Principles of Immunology* (New York: McGraw-Hill, 1957); A. C. Guyton, *Basic Neuroscience: Anatomy and Physiology* (Philadelphia: W. B. Saunders, 1987); B. Kolb and I. Q. Winshaw, *Fundamentals of Human Neuropsychology* 3d ed. (New York: W. H. Freeman, 1990); and R. Dulbecco, *The Origins of Life* (New Haven: Yale University Press, 1987). Specific information on the subject can be found in Niels Jerne, "Antibodies and Learning: Selection versus Instruction," in G. Quarton, T. Melnechuck, and F. O. Scmitt, eds., *The Neurosciences: A Study Program,* vol. 1 (New York: Rockefeller University Press, 1967), 200–205); and D. L. French, R. Laskov, and M. D. Scharff, "The Role of Somatic Hypermutation in the Generation of Antibody Diversity," *Science* 244 (1989): 1152–57.

For more on the theory of evolution from Darwin to today, consult P. Brent,

Charles Darwin: A Man of Enlarged Curiosity (New York: Harper and Row, 1981); R. W. Burkhardt, *The Spirit of System: Lamarck and Evolutionary Biology* (Cambridge, Mass.: Harvard University Press, 1977); R. Dawkins, *The Blind Watchmaker* (New York: W. W. Norton, 1987) and *The Selfish Gene* (Oxford: Oxford University Press, 1989); S. J. Gould and R. C. Lewontin, "The Spandrels of San Marco and the Panglossian Program: A Critique of the Adaptionist Programme," *Proceedings of the Royal Society of London* 205 (1979); M. Konner, *The Tangled Wing: Biological Constraints on the Human Spirit* (New York: Henry Holt & Co., 1982); J. Tooby and L. Cosmides, "From Evolution to Behavior: Evolutionary Psychology as the Missing Link," in J. Dupré, ed., *The Latest on the Best: Essays on Evolution and Optimality* (Cambridge, Mass.: MIT Press, 1987), 277–306; and J. Tooby and L. Cosmides, "On the Universality of Human Nature and the Uniqueness of the Individual: The Role of Genetics and Adaptation," *Journal of Personality* 58 (1990): 17–67.

2. The Plastic Brain and Selection Theory

For an introduction to the developmental side of neurobiology, read K. L. Moore, *The Developing Human* (Philadelphia: W. B. Saunders, 1977). More specific information can be found in D. H. Hubel and T. N. Weisel, "Ordered Arrangement of Orientation Columns in Monkeys Lacking Visual Experience," *Journal of Comparative Neurology* 158 (1974): 307–18; as well as G. M. Innocenti et al., "Maturation of Visual Callosal Connections in Visually Deprived Kittens: A Challenging Critical Period," *Journal of Neuroscience* 5 (1985): 255–67. For a selected review of the work on neural organization, read W. Singer and J. P. Rauschecker, "Central Core Control of Developmental Plasticity in the Kitten Visual Cortex, vol. 2 Electrical Activation of Mesencephalic and Diencephalic Projections," *Experimental Brain Research* 41 (1982): 199–215; W. Singer, "Activity-Dependent Self-organization of Synaptic Connections as a Substrate of Learning," in J. P. Changeux and M. Konishi, *The Neural and Molecular Bases of Learning* (Dahlem Konferenzen, Chichester: John Wiley & Sons, 1987), 301–36; and C. Von der Malsburg and W. Singer, "Principles of Cortical Network Organization," in P. Rakic and W. Singer, eds., *Neurobiology of Neocortex* (New York: John Wiley & Sons, 1988), 69–99. For a more general view of neurobiology, an excellent collection of some current perspectives on the neurosciences can be found in C. B. Trevarthen, ed., *Brain Circuits and Functions of the Mind: Essays in Honor of Roger W. Sperry* (Cambridge: Cambridge University Press, 1990); and Niels Jerne, "Antibodies and Learning: Selection versus Instruction," in G. Quarton, T. Melnechuck, and F. O. Schmitt, eds., *The Neurosciences: A Study Program*, Vol. 1 (New York: Rockefeller University Press, 1967).

3. The Developing Mind

An eloquent discussion of developmental biology is provided by S. J. Gould in *Ontogeny and Phylogeny* (Cambridge, Mass.: Harvard University Press, 1977). Articles on the question of genetics and the environment include R. McCall, "Nature-Nurture and the Two Realms of Development," *Child Development* 52 (1981): 1–12; H. H. Goldsmith, "Genetic Influences on Personality from Infancy to Adulthood," *Child Development* 54 (1983): 331–55; and D. C. Rowe and R. Plomin, "The Importance of Nonshared (E_1) Environmental Influences in Behavioral Development," *Developmental Psychology* 17 (5 [1981]): 517–31. Concepts in language development are spelled out by Noam Chomsky in *Rules and Representations* (New York: Columbia University Press, 1980). Other readings on language and category development include Massimo Piattelli-Palmarini, "Evolution, Selection and Cognition: From 'Learning' to Parameter Setting in Biology and in the Study of Language," *Cognition* 31 (1989): 1–44; E. Dromi, *Early Lexical Development* (Cambridge: Cambridge University Press, 1987); S. Carey, "The Child as Word Learner," in M. Halle et al., *Linguistic Theory and Psychological Reality* (Cambridge, Mass.: MIT Press, 1978), 264–93; and S. A. Gelman and E. M. Markman, "Categories and Induction in Young Children," *Cognition* 23 (1986): 183–209.

For a recent discussion of sensitive periods, consult Marc Bornstein, "Sensitive Periods in Development: Structural Characteristics and Causal Interpretations," *Psychological Bulletin* 105 (2 [1989]): 179–97; P. Marler, "A Comparative Approach to Vocal Learning: Song Development in White-Crowned Sparrows," *Journal of Comparative and Physiological Psychology* 71 (1970): 1–25; and R. N. Aslin, "Experimental Influences and Sensitive Periods in Perceptual Development: A Unified Model," in R. N. Aslin, J. R. Alberts, and M. R. Peterson, eds., *Development of Perception: Psychobiological Perspectives,* vol. 2 (New York: Academic Press, 1981), 45–93. Articles that provide a thorough methodological and theoretical basis for developmental constraints include D. G. Freedman, "Infancy, Biology and Culture," in L. P. Lipsitt, ed., *Developmental Psychobiology: The Significance of Infancy* (Hillsdale, N.J.: Erlbaum, 1976); Frank Keil, "Constraints on Constraints: Surveying the Epigenetic Landscape," and Ellen Markman, "Constraints Children Place on Word Meanings," *Cognitive Science* 14 (1990): 135–68 and 57–77, respectively. The topic of personality development is broadly addressed in Jerome Kagan et al., *Infancy: Its Place in Human Development* (Cambridge, Mass.: Harvard University Press, 1978); and A. H. Buss and R. Plomin, *Temperament Theory of Personality Development* (New York: John Wiley & Sons, 1975). For more on the philosophy of these arguments, consult W. V. Quine, *Philosophy of Logic,"* and *"Pursuit of Truth* (Cambridge, Mass.: Harvard University Press, 1986 and 1990, respectively).

4. Language and Selection Theory

A number of arguments in chapter 4 are supported by the recent work of Steven Pinker and Paul Bloom. See "Natural Language and Natural Selection," *Behavioral and Brain Sciences* 13 (4 [1990]): 723–24. I have illustrated many of the chapter's arguments with references to their work and, wherever appropriate, included examples of other researchers cited in Pinker and Bloom. A few of these include E. Bates, D. Thal, and V. Marchman, "Symbols and Syntax: A Darwinian Approach to Language Development," in N. Krasnegor, D. Rumbaugh, M. Studdert-Kennedy, and R. Schiefelbusch, eds., *The Biological Foundations of Language Development* (Oxford: Oxford University Press, 1989) (p. 728 of P&B); S. J. Gould, "Panselectionist Pitfalls in Parker and Gibson's Model of the Evolution of Intelligence," *The Behavioral and Brain Sciences* 2 (1979): 385–86 (p. 720 of P&B); and D. Premack, " 'Gavagai!' or the Future History of the Animal Language Controversy," *Cognition* 19 (1985): 207–96 (pp. 723–24 of P&B).

Other material by and about two of the architects of current language theory include Noam Chomsky, *Knowledge of Language: Its Nature, Origin, and Use* (New York: Praeger, 1986), *Reflections on Language* (New York: Pantheon Books, 1975), and *Language and Mind* (New York: Harcourt Brace Jovanovich, 1972); and Steven Pinker, "Resolving a Learnability Paradox in the Acquisition of the Verb Lexicon," in M. L. Rice and R. L. Schiefelbusch, eds., *The Teachability of Language* (Baltimore: P. H. Brookes, 1989), *Language Learnability and Language Development* (Cambridge, Mass.: Harvard University Press, 1984), "Formal Models of Language Learning," *Cognition* 7 (1979): 217–83, and "Productivity and Conservatism in Language Acquisition," in W. Demopoulos and A. Marras, eds., *Language Learning and Concept Acquisition: Foundational Issues* (Norwood, N.J.: Ablex, 1986). See also R. P. Botha, *Challenging Chomsky* (New York: Basil Blackwell, 1989); and J. Leiber, *Noam Chomsky: A Philosophic Overview* (Boston: Twayne, 1975). For some specific articles on the biology of language read M. S. Gazzaniga, "Organization of the Human Brain," *Science* 245 (1989): 947–952; M. Gopnik, "Genetic Basis of Grammar Defect" (*Nature* 6 (1990): 26; and M. T. Ghishelin, "Evolutionary Anatomy and Language," *Behavioral and Brain Sciences* 3 (1980): 20–21. For more on the sociological and philosophical contributions to the argument consult R. B. Lee, *The !Dobe Kung* (New York: Holt, Rinehart and Winston, 1984); D. C. Dennett, "Passing the Buck to Biology," and A. Morton, "There are Many Modular Theories of Mind," both in *The Behavioral and Brain Sciences* 3 (1980): 19 and 29, respectively.

5. Specialized Brain Circuits

A thorough account of the history of IQ testing and society's conception of intelligence can be found in Howard G. Gardner, *Frames of Mind* (New York: Basic Books, 1983); and Robert Sternberg, *Beyond IQ* (Cambridge: Cambridge University Press, 1985). Perspectives on the question of intelligence in other species are discussed in D. Premack and A. Premack, *The Mind of the Ape* (New York: W. W. Norton, 1983); and A. Premack, *Why Chimps Can Read* (New York: Harper and Row, 1976). For a discussion of intelligence viewed through an evolutionary lens, consult J. C. Eccles, *Evolution of the Brain: Creation of the Self* (London: Routledge, 1989); Sandra Scarr, "An Evolutionary Perspective on Infant Intelligence: Species Patterns and Individual Variations," in M. Lewis, ed., *Origins of Intelligence* (New York: Plenum Press, 1983), 191–224; and G. S. Omenn and A. G. Motulsky, "Biochemical Genetics and the Evolution of Human Behavior," in L. Ehrman, G. S. Omenn, and E. Caspari, eds., *Genetics, Environment and Behavior* (New York: Academic Press, 1972). For a humbling account of what we can't learn, read Derek Bickerton, *Language and Species* (Chicago: University of Chicago Press, 1990), and *Dynamics of a Creole System* (New York: Cambridge University Press, 1975).

6. Selecting for Mind

For a more detailed discussion of the interpreter and other facets of split-brain research, read Michael Gazzaniga, *The Integrated Mind* (New York: Plenum Press, 1978) and "Cognitive and Neurologic Aspects of Hemispheric Disconnection in the Human Brain," *Discussions in Neurosciences, FESN* (1987): 1–68. A fascinating collection of essays on the mechanics of thought and beliefs, including A. Leslie "The Necessity of Illusion: Perception and Thought in Infancy," can be found in Larry Weiskrantz, ed., *Thought Without Language* (New York: Oxford University Press, 1988). For specific studies on the development of the perceptual and cognitive tools of belief systems, see R. Baillargeon, E. S. Spelke, and S. Wasserman, "Object Permanence in Five-month-old Infants," *Cognition* 20 (1985): 191–208; P. J. Kellman, H. Gleitman, and E. S. Spelke, "Object and Observer Motion in the Perception of Objects by Infants," in *Journal of Experimental Psychology: Human Perception and Performance* 13 (1987): 586–93; A. S. Steri and E. S. Spelke, "Haptic Perception of Objects in Infancy," in *Cognitive Psychology* 20 (1 [1988]): 1–23; and P. J. Kellman and E. S. Spelke, "Perception of Partly Occluded Objects in Infancy," in *Cognitive Psychology* 15 (1983): 483–524.

For a biological approach to beliefs, read J. C. Eccles, *Evolution of the Brain:*

Creation of the Self (London: Routledge, 1989); and R. L. Trivers, "The Evolution of Reciprocal Altruism," *The Quarterly Review of Biology* 46 (1971): 35–57. Finally, to review the evidence for beliefs from social psychology, consult Richard Nisbett and Lee Ross, *Human Inference: Strategies and Shortcomings of Social Judgment* (Englewood Cliffs, N.J.: Prentice-Hall, 1980); C. Lord, L. Ross, and M. R. Lepper, "Biased Assimilation and Attitude Polarization: The Effects of Prior Theories on Subsequently Considered Evidence," *Journal of Personality and Social Psychology* 37 (11 [1979]): 2098–2109; R. E. Nisbett and T. D. Wilson, "The Halo Effect: Evidence for Unconscious Alteration of Judgments," *Journal of Personality and Social Psychology* 35 (1977): 250–56; G. R. Salancik, "Extrinsic Attribution and the use of Behavioral Information To Infer Attitudes," *Journal of Personality and Social Psychology* 34 (1976): 1302–12; Amos Tversky and D. Kahneman, "Availability: A Heuristic for Judging Frequency and Probability," *Cognitive Psychology* 5 (1973): 207–32; and Josef Perner, *Understanding the Representational Mind* (Cambridge, Mass.: MIT Press, 1991).

7. Addictions, Compulsions, and Selection Theory

There is no shortage of literature on the subject of addiction. For a general review of some current thoughts on alcoholism and other addictions, read J. Oxford, *Excessive Appetites: A Psychological View of Addictions* (New York: John Wiley & Sons, 1985); S. Peele, *The Meaning of Addiction* (Lexington, Mass.: D. C. Heath and Co., 1985); G. A. Marlatt, "Alcohol, the Magic Elixir: Stress, Expectancy, and the Transformation of Emotional States," in E. Gottheil, K. A. Druly, S. Pashko, and S. P. Weinstein, eds., *Stress and Addiction* (New York: Brunner/Mazel, 1987); and T. K. Li and J. C. Lockmuller, "Why Are Some People More Susceptible to Alcoholism?," *Alcohol Health and Research World* 13 (1989): 310–15. For some hard facts on the statistics behind drug abuse, consult *Seventh Special Report to the U.S. Congress on Alcohol and Health* (U.S. Department of Health and Human Services, 1990); *National Institute on Drug Abuse: Overview of the 1990 National Household Survey on Drug Abuse;* M. E. Hilton, "Drinking Patterns and Drinking Problems in 1984: Results from a General Population Survey," *Alcoholism* (NY) 11 (1987): 167–75; C. J. P. Eriksson, "Finnish Selection Studies on Alcohol-Related Behaviors," in G. E. McClearn et al., eds., *Development of Animal Models as Pharmacogenetic Tools* (Washington, D.C.: Supt. of Docs., U.S. Govt. Print. Off., 1981), 119–45; L. N. Robins, J. E. Helzer, and D. H. Davis, "Narcotic Use in Southeast Asia and Afterward," *Archives of General Psychiatry* 32 (1975): 955–61; and S. Edwards, "A Sex Addict Speaks," *SIECUS Reports* 14 (1986).

For a discussion of the genetic factors involved with addictions, read C. Holden, "Probing the Complex Genetics of Alcoholism," *Science* 251 (1991): 163–64; D. W. Goodwin, "Genetic Influences in Alcoholism," *Advances in Inter-*

nal Medicine 32 (1987): 283–98; B. Tabakoff and P. L. Hoffman, "Genetics and Biological Markers of Risk for Alcoholism," *Public Health Rep* 103 (6 [1988]): 690–98; C. R. Cloninger, M. Bohman, and S. Sigvardsson, "Inheritance of Alcohol Abuse," *Archives of General Psychiatry* 38 (1981): 861–68; and D. W. Goodwin, F. Schulsinger, N. Moller, L. Hermansen, G. Winokur, and S. Guze, "Drinking Problems in Adopted and Nonadopted Sons of Alcoholics," *Archives of General Psychiatry* 31 (1974): 164–69. For a discussion of some more specific biological links to addiction, read J. R. Cooper, F. E. Bloom, and R. H. Roth, *The Biochemical Basis of Neuropharmacology* 4th ed. (Oxford: Oxford University Press, 1982); V. E. Pollock et al., "The EEG after Alcohol Administration in Men at Risk for Alcoholism," *Archives of General Psychiatry* 40 (1983): 857–61; as well as H. Begleiter et al., "Event-Related Potentials in Boys at Risk for Alcoholism," *Science* 225 (1984): 1493–96. Also see E. P. Noble, K. Blum, T. Ritchie, A. Montgomery, and P. Sheridan, "Allelic Association of the D_2 Dopamine Receptor Gene with Receptor-Binding Characteristics in Alcoholism," in *Archives of General Psychiatry* 48 (1991): 648–54; G. J. Gatto, J. M. Murphy, M. B. Waller, W. J. McBride, L. Lumeng, and T. K. Li, "Persistence of Tolerance to a Single Dose of Ethanol in the Selectively Bred Alcohol-Preferring P Rat," *Pharmacology, Biochemistry, and Behavior* 28 (1 [1987]): 105–10; and G. F. Koob and F. E. Bloom, "Cellular and Molecular Mechanisms of Drug Dependence," *Science* 242 (1988): 715–23.

Finally, for a discussion of possible links between addictions and other psychiatric or sociological disorders, consult M. Bohman, "Some Genetic Aspects of Alcoholism and Criminality: A Population of Adoptees," *Archives of General Psychiatry* 35 (1978): 269–76; Z. Hrubec and G. S. Omenn, "Evidence of Genetic Predisposition to Alcoholic Cirrhosis and Psychosis," *Alcoholism: Clinical and Experimental Research* 5 (1981): 207–15; R. A. McCormick, J. Taber, N. Kruedelbach, and A. Russo, "Personality Profiles of Hospitalized Pathological Gamblers: The California Personality Inventory," *Journal of Clinical Psychology* 43 (5 [1987]): 521–27; and C. R. Cloninger, T. Riech, and S. Yokoyama, "Genetic Diversity, Genome Organization, and Investigation of the Etiology of Psychiatric Diseases," in *Psychiatric Developments* 3 (1983): 225–46.

8. Selection Theory and the Death of Psychoanalysis

For background into the life and writings of Freud, read, J. Strachey, ed., "The Complete Psychological Works of Sigmund Freud," in *Melancholia,* vol. 1 (London: Hogarth Press, 1865); and E. Freud, L. Freud, and I. Grubrich-Simitis, eds., *Sigmund Freud: His Life in Pictures and Words* (New York: W. W. Norton, 1978). For material on the topic of psychoanalytic theory, consult Sandor Ferenczi, "Psycho-Analytical Observations on Tic," *International Journal of Psychoanalysis* 2

(1921): 1–30; Melford Spiro, *Oedipus in the Trobriands* (Chicago: University of Chicago Press, 1982); and S. Frosh, *Psychoanalysis and Psychology: Minding the Gap* (New York: New York University Press, 1989). And for more on the topic of family violence, don't miss M. Daly and M. Wilson, *Homicide* (New York: Aldine de Gruyter, 1988). Other books on the subject include R. J. Gelles and M. A. Strauss, *Intimate Violence* (New York: Simon and Schuster, 1988); and *Physical Violence in American Families: Risk Factors and Adaptations to Violence in 8,145 Families* (New Brunswick, N.J.: Transaction Publishers, 1990).

On the subject of affective disorders, a variety of perspectives can be seen in J. H. Boyd and M. M. Weissman, "Epidemiology of Affective Disorders: A Re-examination and Future Directions," *Archives of General Psychology* 38 (1981): 1039–45; G. Carey, "Big Genes, Little Genes, Affective Disorder, and Anxiety," *Archives of General Psychiatry* 44 (1987): 486–91; D. S. Janowsky and D. H. Overstreet, "Cholinergic Dysfunction in Depression," in A. Geisler, A. Mork, and R. Klysner, eds., *Neurochemical Correlates of Affective Disorders and Their Treatment: A Satellite Symposium Affiliated to the XVIth C.I.N.P. Congress* (Copenhagen: University of Copenhagen, 1988); A. J. Prange, et al., "L-Tryptophan in Mania-Contribution to a Permissive Hypothesis of Affective Disorders," *Archives of General Psychiatry* 30 (1974): 56–62; and F. K. Goodwin and K. R. Jamison, "The Natural Course of the Manic-Depressive Illness," in R. Post and J. Ballenger, eds., *Neurobiology of Mood Disorders* (Baltimore: Williams and Wilkins, 1984). Current treatments of affective disorders are discussed in G. Murphy et al., "Cognitive Therapy and Pharmacotherapy," *Archives of General Psychiatry* 41 (1983): 33–41; and I. M. Blackburn et al., "The Efficacy of Cognitive Therapy in Depression: A Treatment Trial Using Cognitive Therapy and Pharmacotherapy, Each Alone and in Combination," *British Journal of Psychiatry* 139 (1981): 181–89. Neurobiological models of psychiatric disorders can be found in W. E. Rinn, "The Neuropsychology of Facial Expression," *Psychological Bulletin* 95 (1984): 52–77; H. F. Harlow, "Age-Mate or Peer Affectional System," *Advances in the Study of Behavior* 2 (1969): 333–83; H. F. Harlow and C. Mears, *The Human Model: Primate Perspectives* (New York: V. H. Winston & Sons, 1979); J. Tooby and L. Cosmides, "On the Universality of Human Nature and the Uniqueness of the Individual: The Role of Genetics and Adaptation," *Journal of Personality* 58 (1990): 17–67; R. J. Gelles and M. A. Strauss, "Family Experience and Public Support for the Death Penalty," in R. J. Gelles, ed., *Family Violence* (Beverly Hills, Calif.: Sage, 1979).

9. Health Care, Aging, and Selection Theory

For more information on the application of evolutionary theory to medicine and other aspects of life, read G. C. Williams and R. M. Nesse, "The Dawn of

Darwinian Medicine," *The Quarterly Review of Biology* 66 (1991): 1–32; and R. Trivers, *Social Evolution* (Menlo Park, Calif.: Benjamin/Cummings, 1985). There are a number of recent contributions to the question of how stress and immunity are related. A few of these include S. Cohen, D. A. G. Tyrrell, and A. P. Smith, "Psychological Stress and Susceptibility to the Common Cold," *New England Journal of Medicine* 325 (1991): 606; R. Sapolsky, M. Armanini, D. Packan, and G. Tombaugh, "Stress and Glucocorticoids in Aging," *Endocrinology and Metabolism Clinics of North America* 16 (4 [1987]): 965–80; A. Breier, M. Albus, D. Pickar, T. P. Zahn, O. M. Wolkowitz, and S. M. Paul, "Controllable and Uncontrollable Stress in Humans: Alterations in Mood and Neuroendocrine and Psychophysiological Function," *American Journal of Psychiatry* 144 (1987): 1419–25; and J. K. Kiecolt-Glaser, R. Glaser, E. C. Shuttleworth, C. S. Dyer, P. Ogrocki, and C. E. Speicher, "Chronic Stress and Immunity in Family Caregivers of Alzheimer's Disease Victims," *Psychosomatic Medicine* 49 (1987): 523–35.

For an understanding of the chemical basis of immunology, consult E. Gould, C. S. Woolley, and B. S. McEwen, "Naturally Occurring Cell Death in the Developing Dentate Gyrus of the Rat," *The Journal of Comparative Neurology* 304 (1991): 408–18; P. Medawar, "The Nobel Lectures in Immunology," *Scandinavian Journal of Immunology* 33 (4 [1991]): 337–44; L. Jacobson and R. Sapolsky, "The Role of the Hippocampus in Feedback Regulation of the Hypothalamic-Pituitary Adrenocortical Axis," *Endocrine Reviews* 12 (2 [1991]): 118–34; M. Profet, "The Function of Allergy: Immunological Defense against Toxins," *The Quarterly Review of Biology* 66 (1 [1991]): 23–62; M. L. Laudenslager, S. M. Ryan, R. C. Drugan, R. L. Hyson, and S. F. Maier, "Coping and Immunosuppression: Inescapable but not Escapable Shock Suppresses Lymphocyte Proliferation," *Science* 221 (1983): 568–70; S. J. Schleifer, S. E. Keller, R. N. Bond, J. Cohen, and M. Stein, "Major Depressive Disorder and Immunity," *Archives of General Psychiatry* 46 (1989): 81–87; J. E. Blalock, D. Harbour-McMenamin, and E. M. Smith, "Peptide Hormones Shared by the Neuroendocrine and Immunologic Systems," *Journal of Immunology* 135 (1985): 858s–61s; and M. L. Laudenslager, M. Reite, and R. J. Harbeck, "Suppressed Immune Response in Infant Monkeys Associated with Maternal Separation," *Behavioral and Neural Biology* 36 (1982): 40–48.

On the issue of control and helplessness, further reading can be found in the classic S. F. Maier and M. E. P. Seligman, "Learned Helplessness: Theory and Evidence," *Journal of Experimental Psychology: General* 105 (1976): 3–47; S. Nolen-Hoeksema, J. Girgus, and M. E. P. Seligman, "Learned Helplessness in Children: A Longitudinal Study of Depression, Achievement, and Explanatory Style," *Journal of Personality and Social Psychology* 51 (1986): 435–42; C. Peterson, M. E. P. Seligman, and G. E. Vaillant, "Pessimistic Explanatory Style is a Risk Factor for Physical Illness: A Thirty-five-year Longitudinal Study," *Journal of Personality and Social Psychology* 55 (1988): 23–27; J. Rodin, "Aging and Health: Effects of the Sense of Control," *Science* 233 (1986): 1271–76; D. S. Kirschenbaum, J. Sherman, and J. D. Penrod, "Promoting Self-Directed Hemodialysis: Measurement and Cognitive-Behavioral Intervention," *Health Psychology* 6 (1987): 373–85; E. J. Langer and J. Rodin, "The Effects of Choice and Enhanced Personal Responsibility for the Aged: A Field Experiment in an Institutional Setting," *Journal of Personality and Social Psychology* 34 (1976): 191–198; and G. Affleck, H. Tennen, C. Pfeifer, and J. Fifield, "Appraisals of Control and Predictability in Adapting to a Chronic Disease," *Journal of Personality and Social Psychology* 53 (1987): 273–79.

Finally, for more on the effects of social support systems, read L. T. Empey and M. L. Ericson, *The Provo Experiment: Evaluating Community Control of Delinquency* (Lexington, Mass.: Lexington Books, 1972); M. M. Craig and P. W. Furst, "What Happens after Treatment? A Study of Potentially Delinquent Boys" in *Social Service Review* 39 (1965): 165–71; G. R. Robin, "Anti-Poverty Programs and Delinquency," *Journal of Criminal Law, Criminology, and Police Science* 60 (1969): 323–31, W. B. Miller, "The Impact of a "Total Community" Delinquency Control Project," *Social Problems* 10 (1962): 168–91; S. Cohen, "Psychosocial Models of the Role of Social Support in the Etiology of Physical Disease," *Health Psychology* 7 (1988): 269–97; M. Pilisuk, R. Boylan, and C. Acredolo, "Social Support, Life Stress, and Subsequent Medical Care Utilization," *Health Psychology* 6 (1987): 273–88; S. Cohen and T. A. Wills, "Stress, Social Support, and the Buffering Hypothesis," *Psychological Bulletin* 98 (1985): 310–57; and Joan McCord, "A Thirty-year Follow-up Study of Treatment Effects," *American Psychologist* 33 (1978): 284–89.

INDEX